Electrical engineering without prior knowledge

Understand the basics within 7 days

Benjamin Spahic

Table of contents

1 Preface and introduction 1
2 The fundamentals of mathematics 4
2.1 Solving equations 4
2.2 Exponential functions 5
2.3 Powe laws 5
2.4 The Euler's number e 6
2.5 Logarithms 7
2.6 Logarithms table 8
2.7 The Greek alphabet 8
2.8 Sine, cosine, tangent 10
2.9 Sine and cosine functions 11
2.10 Arc sine, arc cosine, arc tangent 12
2.11 The two-dimensional coordinate system 13
3 Physics basics 16
3.1 Notation, upper case letters, lower case letters 16
3.2 Prefixes for a wide dynamic range 17
3.3 The "Système international d'unités" 19
3.4 Derived SI units 20
3.5 Displaying differences 20
3.6 Conservation of energy and efficiency 21
3.7 Energy 23
3.8 Power 23
4 From water model to circuit 26
4.1 Atoms, electrons, protons 27
4.2 When does material conduct electricity? 28
5 The electric field 30

5.1	Representation of E-fields	30
5.2	The force in the electric field	32
5.3	Electric potential and voltage	34
5.4	Current	36
5.5	Technical and physical current direction	37
6	The magnetic field	38
6.1	Elementary magnets	39
6.2	Displaying magnetic fields	41
6.3	Electromagnetism	42
6.4	Induction law	43
6.5	Magnetic flux and induction	44
6.6	The Lenz Rule	45
6.7	The Lorentz force	46
6.8	The three-finger rule	47
6.9	Overview: E-field and B-field	48
7	Markings and treasure symbols	50
7.1	Mass and earth	50
7.2	The consumer	50
7.3	The completed circuit	51
7.4	What happens without consumers?	52
7.5	Counting arrow systems	52
7.6	Voltage arrows	52
7.7	Current arrows	53
7.8	Producer and consumer arrow system	54
7.9	Kirchhoff's laws	54

7.10	Kirchoff law for nodes	55
7.11	Kirchoff law for meshes	56
8	Electrical resistance	58
8.1	Series connection of resistors	61
8.2	Voltage divider	61
8.3	Parallel connection of resistors	62
8.4	Special form for two resistors	63
8.5	Current divider	63
8.6	Electrical power	64
8.7	Application example: Resistors in a power supply unit	65
9	Semiconductors: PN junctions, diodes, transistors	66
9.1	Structure of a diode	67
9.2	Excursus: LED	70
9.3	The transistor	71
9.4	The bipolar transistor	72
9.5	The field effect transistor	74
10	The capacitor	78
10.1	Charging a capacitor	81
10.2	Discharging the capacitor	85
10.3	How much energy can a capacitor store?	87
10.4	Application area of capacitors	88
11	The coil	90
11.1	Magnetic coupling	92
11.2	Switch-on process of a coil	93
11.3	Switching off a coil	96
11.4	How much energy can a coil store?	98

11.5	Comparison capacitor and coil	99
12	Practical example - LED switch-on delay	100
12.1	The circuit	101
12.2	Calculating the time delay	102
13	Introduction to alternating current theory	104
13.1	Power generation	104
13.2	Power generation by means of generators	108
13.3	Structure of the power grid	119
14	Components in the AC circuit	128
14.1	The resistance	129
14.2	The capacitor	130
14.3	The coil	134
14.4	Active, reactive and apparent power	138
14.5	The electromagnetic oscillating circuit	142
14.6	Electromagnetic radiation	148
15	Conclusion	152
	Free eBook	153

Imprint

PBD  Verlag

Author: Benjamin Spahic
Addresse:
Konradin-Kreutzer-Str. 12
76684 Oestringen
Germany

Proofreading/ Editing: Oliver Nova
Cover: Kim Nusko
Proofreader Second Version: Roland Bümel and Mentorium GmbH
ISBN: 9798687840071
ISBN Hardcover: 979-8352008287
Email: BenjaminSpahic@pbd-verlag.de.de
Facebook: Benjamin Spahic
Electrical engineering without prior knowledge
First publication 23.05.2020
Distribution through kindledirectpublishing
Amazon Media EU S.à r.l., 5 Rue Plaetis, L-2338, Luxembourg

Disclaimer: The author accepts no responsibility for the topicality, correctness, completeness or quality of the information provided. Liability claims against the author relating to material or immaterial damage caused by the use of the information provided or by the use of incorrect or incomplete information are excluded to the fullest extent permissible.

Furthermore, no guarantee can be given for the achievement of the described skills

1 Preface and introduction

Hardly any other subject area is as diverse as electrical engineering or has as much control over our everyday life.

In the morning, we are woken up by the smartphone or the digital alarm clock. Without integrated circuits and synchronized clocks, a large part of the population would probably not get out of bed on time. Then we get up and switch on the light as a matter of course – without electricity, we would have to grope our way through the corridors by candlelight to find our way to the kitchen.

During breakfast, we check our e-mails or read news online – without digital data transmission we would not be aware of what was new in our world.

"The progress of technology is based on adapting it so that you don't even really notice it's part of your daily life."

- Bill Gates

This procedure runs through our entire everyday life. Thanks to the control electronics in the car, our vehicle brings us safely to work, machines and computers ensure an ever-increasing economic performance and, at the end of the day when we lie relaxed on the couch, we can enjoy the latest Netflix series or watch funny cat videos on YouTube. Electrical engineering is the foundation for all these areas, from the generation and provision of the power grid to data processing and transmission and to nanotechnology.

However, for all its importance, there is one big problem: enthusiasm for understanding and learning electrical engineering is very limited in society. Only a small elite is interested in the subject area.

Since you bought this book, you seem to belong to this circle. Maybe you are still a student thinking about studying engineering, maybe you are a career changer who just wants to understand the basics, or maybe you are a programmer who wants to learn more about hardware. Either way, you will not regret dealing with the matter.

If you are dealing with electrical engineering for the first time, you will find various books, some of which have over 500 pages and are completely unsuitable for newcomers: pages and pages of mathematical derivations, which you forget after a week. Of course, these books also have their right to exist, especially if you want to question and understand a matter in the smallest detail, but for the majority of interested people, this is neither necessary nor effective.

And it is precisely because of this problem that this book has emerged.

It is a beginner's guide for those who are eager to learn and understand the basic principles of electrical engineering without much previous knowledge.

What is tension? How do I calculate my electricity consumption, and how can I build up a small electrical circuit myself? This book attaches great importance to using real values and examples and not utopian examples and esoteric calculations. This book offers practical relevance and at the same time illuminates the basic mathematical principles and derivations as far as necessary. After you have read this beginner's guide, you will have a feel for electrical quantities. You will be able to classify numbers correctly in a factual context and know what is important.

Requirements and level of knowledge:

This book is suitable for anyone with a basic enthusiasm for technology and mathematical understanding. Since for some readers it may be a little while since their last mathematics or physics lesson, the first chapter deals with the basics of mathematics and physics. It is therefore assumed that the reader has a technical understanding, but no in-depth previous knowledge.

If you can say to yourself that you do not have any catching-up to do in these areas, you can start, if necessary, with the third chapter, in which the analogy of the electricity and water cycle is presented. However, it is advisable to at least skim the basics again.

In the book, you will find the following icons in certain places:

Calculation symbols: It gets more complex here. An excursus or mathematical derivation is cited.

The derivation of a topic area is helpful for understanding but is not essential and is rather intended for reference.

Light bulb: Here, the key points of a chapter are summarised. These statements are good for reference or when you skim over a topic area again.

Attention: Frequent errors are mentioned here. It is shown where and why one often encounters obstacles or false assumptions.

Pocket calculator: Sample calculations or comprehension questions to understand and internalise.

Newly acquired knowledge is retained much better if it is applied immediately. If you have to brood over a question of understanding, this is an indication that you should go through the previous chapter again before continuing.

Now, I wish you a lot of fun while reading and diving into the wonderful world of electrical engineering.

2 The fundamentals of mathematics

When you dive into electrical engineering, juggling with terms and equations becomes the order of the day. Mathematics provides us with the basis for this. It serves us as a tool.

Just as a carpenter needs to know how to use a hammer and chisel, we need to know how to properly summarise or simplify formulas. Basic arithmetic laws, function types and number systems are covered below. Those who have obtained a university entrance qualification will already be familiar with most of the areas, but partial aspects are also discussed that one only learns in technical high schools, for example. From experience, mathematics is a necessary evil, so each subject area is only dealt with as far as it is important for the understanding of this book.

2.1 Solving equations

The aim of solving an equation is to rearrange the equation so that we end up with the variable we are looking for on one side of the equal's sign.

$3x + 8 = -2x + 3$
...
$x = -1$

To do this, we need to edit the equation in several steps to isolate the variable.

 When solving an equation, you transform it step by step until the variable you are looking for (e.g., x) is alone and positive on one side. The transformations are called equivalent transformations. This does not falsify the statement of the equation.

For example, we can add or subtract a constant or variable on both sides of an equation, or multiply, divide, exponentiate etc. both sides by a factor. When applying an equivalence transformation, write it at the end of the line together with a vertical line.

$$3x + 8 = -2x + 3 \quad | + 2x$$
$$5x + 8 = 3 \quad | - 8$$
$$5x = -5 \quad |:5$$
$$x = -1$$

All transformations must always take place on both sides of the equation. We will encounter the transformation of equations several times in each chapter.

2.2 Exponential functions

Exponential functions occur more often in everyday life than we think. Almost every natural process can be traced back to an exponential function: the growth of bacteria, the heating or cooling of any matter (whether food, sand or metal) or electrotechnical processes such as the charging and discharging of accumulators, battery storage or capacitors. In order to understand how these processes work, we first turn to the mathematical basics – the exponential functions.

An exponential function is a function of the form

$$f(x) = a^x$$

Here, a is called the base and x the exponent (colloquially high number). The base must be a real number that is greater than 0 and not equal to 1. The exponent is usually part of the real numbers. Note also the case for $x = 0$.

$$a^0 = 1$$

For any base a.

2.3 Power laws

Power laws are applicable to terms with similar properties and allow us to summarise powers more clearly. In electrical engineering, you have to calculate a lot with exponents, so it helps if you have a few tricks at hand.

All the following equations always work in both directions!

Power with a negative exponent

If the exponent of power is negative, the power can be rewritten as

$$a^{-b} = \frac{1}{a^b}$$

$$2^{-2} = \frac{1}{2^2}$$

Multiplication of powers with the same base

If two or more powers with the same base are multiplied together, the exponents add up. The base remains unchanged.

$$a^b \times a^c = a^{b+c}$$

$$3^2 \times 3^5 = 3^{2+5} = 3^7$$

Division of powers with the same base

If two or more powers with the same base are divided, the exponents subtract. The base remains unchanged. The derivation is obtained by writing the division as multiplication with a negative exponent.

$$\frac{a^b}{a^c} = a^b \times a^{-c} = a^{b-c}$$

$$\frac{2^5}{2^3} = 2^{5-3} = 2^2$$

Multiplication of powers with equal exponents

If two or more powers with the same exponent but different bases are multiplied together, the bases are multiplied. The exponent remains unchanged.

$$a^c \times b^c = (a \times b)^c$$

$$2^5 \times 3^5 = (2 \times 3)^5 = 6^5$$

Division of powers with equal exponents

If two or more powers with the same exponent but different bases are divided, the bases are divided. The exponent remains unchanged.

$$\frac{a^c}{b^c} = \left(\frac{a}{b}\right)^c$$

$$\frac{2^5}{3^5} = \left(\frac{2}{3}\right)^5$$

Exponentiating powers

If a (base with) power is exponentiated, the exponents are multiplied together.

$$(a^b)^c = a^{b \times c}$$

$$(2^3)^5 = 2^{3 \times 5} = 2^{15}$$

2.4 The Euler's number e

Euler's number e is a constant. It was named after the Swiss mathematician Leonhard Euler and is defined in the range of irrational, real numbers by the limit value

$$e = \sum_{k=0}^{\infty} \frac{1}{k!} = 1 + \frac{1}{1} + \frac{1}{1 \times 2} + \frac{1}{1 \times 2 \times 3} + \cdots = 2.718\ldots$$

The fundamentals of mathematics

This definition is not important for electrotechnical understanding, but it is mentioned for the sake of completeness.

In addition to this definition, there are numerous other limit values that are e approximate.

 For the purposes of this book, it is sufficient to keep in mind the approximate numerical value of 2.72...

The number is of great importance in electrical engineering as well as generally in the entire field of calculus and many other subfields of mathematics.

Euler's number occurs in many natural events such as radioactive decay, natural growth or the charging and discharging of electronic components such as capacitors or coils. When Euler's number forms the basis of an exponential function, it is called an e – function with $f(x) = e^x$

 The special feature of the e – function is that the slope at each point corresponds to the function value at that point. In mathematical terms, this means: $f'(x) = f(x)$

2.5 Logarithms

Logarithms occur just as frequently as exponential functions in everyday life, for example in the human ear, in natural decay, pH values or our perception of brightness.

The basic arithmetic operations, i.e., "plus and minus" as well as "times and divided", are known. For every mathematical operation, there is a corresponding inverse function. For example, if you want to reverse an addition, you subtract; a multiplication is reversed by means of division. The logarithm function is used to reverse exponentiation.

For example, solving the equation: $10^x = 1000$ is asked.

To obtain the solution, i.e., our searched variable x, we apply the logarithm function to the base 10, colloquially "we draw the logarithm to the base 10". The number in the logarithm is called the numerus or logarithmand.

$\log_{10}(10^x) = \log_{10}(1000) \rightarrow x = 3$

The base is written as an index to the logarithm.

In other words, the logarithm solves the problem: "To what number do I have to take the base (10 in the example) to get the result (1000)". The answer in the example is three because

$10^3 = 1000$

For each base, there is a corresponding logarithm. Some occur more frequently and have therefore been given their own abbreviation.

2.6 Logarithms table

The following table shows the notation of the logarithms to the base

Base of the logarithm	Notation	Designation
Any number a	$\log_a z$	Logarithm to base a
2	$lbz = \log_2 z$	Logarithm of two
e	$\ln z = \log_e z$	Natural logarithm
10	$\lg z = \log_{10} z$	Decimal logarithm

The natural logarithm is the logarithm most commonly used in mathematics. The logarithm of two is often used in the IT sector, as a computer works digitally, i.e., calculates binary only with ones and zeros.

2.7 The Greek alphabet

In addition to solving equations and the power laws, we often use the Greek alphabet in electrical engineering, using both upper and lower case letters. The names will be repeated in the coming chapters. The Greek alphabet has a similar structure to ours and is therefore easy to understand. We do not have to learn the complete alphabet by heart. The letters we need will be explained in more detail in the coming chapters. Nevertheless, an overview for reference is not amiss and helps when we are looking for the pronunciation or a specific letter.

The following table shows the Greek alphabet, both upper and lower cases.

Capital letter	Lower case	Pronunciation
A	α	Alpha
B	β	Beta

Γ	γ	Gamma
Δ	δ	Delta
E	ε / ϵ	Epsilon
Z	ζ	Zeta
H	η	Eta
Θ	θ ϑ	Theta
I	ι	Iota
K	κ /	Kappa
Λ	λ	Lambda
M	μ	My [mü]
N	ν	Ny [nü]
Ξ	ξ	Xi
O	o	Omicron
Π	π	Pi
P	ρ	Rho
Σ	σ	Sigma
T	τ	Tau
Y	υ	Ypsilon
Φ	φ / ϕ	Phi
X	χ	Chi
Ψ	ψ	Psi
Ω	ω	Omega

The fundamentals of mathematics

2.8 Sine, cosine, tangent

In addition to applying arithmetic laws, we will look at some trigonometry.

Sine, cosine and tangent describe the ratio of the length of two sides within a right-angled triangle.

The triangle consists of two catheti and a hypotenuse. The cathetus that lies at the angle α and the right angle is called the **adjacent side** of α. The side opposite the angle α is called the **opposite side**.

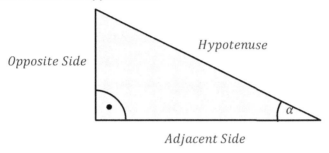

Figure 1: Right-angled triangle

$$\sin \alpha = \frac{\text{Opposite side}}{\text{Hypotenuse}} = \cos(\alpha - 90°)$$

$$\cos \alpha = \frac{\text{Adjacent side}}{\text{Hypotenuse}} = \sin(\alpha + 90°)$$

$$\tan \alpha = \frac{\sin \alpha}{\cos \alpha} = \frac{\text{Opposite side}}{\text{Adjacent side}}$$

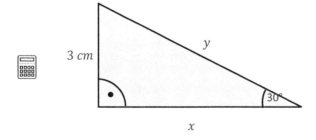

Figure 2: Sine and cosine on the right-angled triangle

$$sin(30°) = \frac{Opposite\ side}{Hypotenuse} = \frac{3\ cm}{y}$$

$$cos(30°) = \frac{Adjacent\ side}{Hypotenuse} = \frac{x}{y}$$

$$tan(30°) = \frac{sin\ \alpha}{cos\ \alpha} = \frac{3\ cm}{x}$$

$$=> y = \frac{3\ cm}{sin(30°)} = \frac{3\ cm}{0.5} = 6\ cm \qquad => x = \frac{3\ cm}{tan(30°)} = \frac{3\ cm}{0.577} \approx 5.2$$

2.9 Sine and cosine functions

If the hypotenuse is set to one in a triangle, the sine of an angle corresponds to its opposite cathetus, the cosine of the angle to its opposite cathetus.

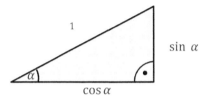

Figure 3: Sine and cosine for hypotenuse of length 1

$$sin\ \alpha = \frac{Opposite\ side}{Hypotenuse} = Opposite\ side\ \ ; cos\ \alpha = \frac{Adjacent\ side}{Hypotenuse} = Adjacent\ side$$

If we then change the angle, α is then changed from 0° to 360°, we obtain a function that expresses the value of the opposite or adjacent side as a function of the angle.

Instead of specifying the angle in degrees, it is common to use a conversion to circle angles or the so-called radians. A circle with the radius $r = 1$ has a circumference of $U = 2\pi$. This circumference is used as a reference for a full angle of 360°. The 360° correspond to 2π, 180° that correspond to π and so on. From the angle, α becomes $x = \frac{\alpha}{360°} \times 2\pi$.

If we plot the length of the sine and the cosine over the angle, we get the sine and cosine function respectively.

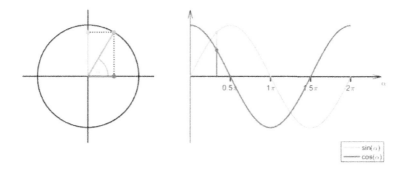

Figure 4: Sine and cosine function

Every natural oscillation consists of superimposed sine and cosine functions.

2.10 Arc sine, arc cosine, arc tangent

The sine, cosine and tangent functions map a ratio or a number onto an angle or a radiant value. Just as the square root is the inverse function of exponentiation or the logarithm function is the inverse function of the exponential function, there are also the corresponding inverse functions for sine, cosine and tangent.

 The arc sine $arcsin()$, arc cosine $arccos()$ and arc tangent $arctan()$ are the inverse functions and allow the radian or angle to be calculated from the ratio value.

In the example, $\sin \alpha = 0.5$ we apply the arc sine to compensate the sine function and get back the corresponding angle.

$\sin \alpha = 0.5$

$\arcsin(\sin \alpha) = \arcsin(0.5)$

$\alpha = \arcsin(0.5) \Rightarrow$ Calculator $\alpha = 30°$

 Often instead of $\arcsin(x)$ the expression $\sin^{-1}(x)$ is used. Analogously $\cos^{-1}(x)$ for the arc cosine or $\tan^{-1}(x)$. Strictly speaking, this is wrong, for example $\sin^{-1}(x) = \frac{1}{\sin(x)} \neq \arcsin(x)$.

This does not correspond to the arc sine. However, the expressions $\sin^{-1}(x)$, $\cos^{-1}(x)$, and $\tan^{-1}(x)$ are widely used and anyone familiar with the subject knows that the arc functions are meant.

2.11 The two-dimensional coordinate system

Before we can conclude the chapter on mathematics, we will look at the representation of numbers and functions in coordinate systems. We will use the Cartesian coordinate system. Most people remember this from school. Cartesian means that the axes are perpendicular to each other.

For the purpose of this book, we will limit ourselves to two dimensions with two axes. The horizontal axis is called the abscissa axis or, more simply, the X-axis. The vertical axis is called the ordinate axis, the vertical axis or simply the Y-axis. We will not consider spatial depth, which isa third dimension, otherwise it can quickly become too complex. The calculations are analogous for both axes. We can enter points in this coordinate system. A point in the mathematical sense is a circle with an infinitely small radius. A point is usually represented as a cross, rectangle or circle. A point has an X and a Y coordinate.

$$P = (x|y)$$

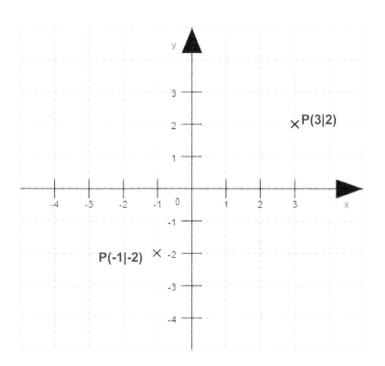

Figure 5: Cartesian coordinate system

The fundamentals of mathematics

! It is important to understand that a coordinate system always refers to an origin or the zero point, which we can determine ourselves!

The zero point always has the coordinates (0|0). It can be the corner of a room, the starting point of a race track or, as on the world map, our poles. Most of the time it results from a task.

💡 By cleverly choosing the zero point, the subsequent calculations can often be simplified.

The great advantage of coordinate systems is that we can represent mathematical facts graphically. This gives us a clearer picture and facilitates understanding.

In addition to individual points, we can also graphically represent entire functions in a coordinate system. The function assigns a y-value to each x-value. An infinite number of values results in a continuous line, the graph of the function.

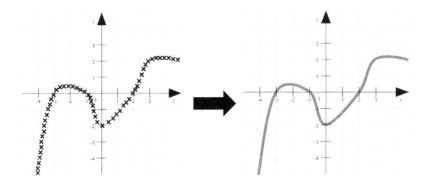

Figure 6: Points in the coordinate system become the function graph

This concludes our brief review of the Cartesian coordinate system.

There are many more coordinate systems than you might think. For example, the position of a point (in relation to the origin) can be described not only as length (X-axis) and height (Y-axis) but also as a radius from the origin and an angle. However, these are not relevant for this book and will therefore not be discussed further.

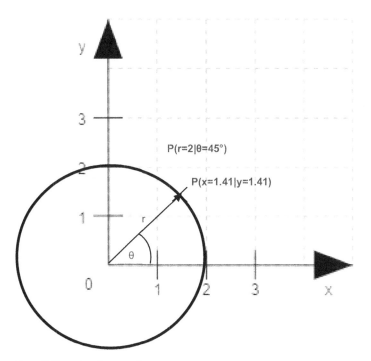

Figure 7: Illustration of two coordinate systems

3 Physics basics

After having struggled through the fundamentals of mathematics, in this chapter, we will look at some conventions of basic physics.

German engineers are known for their love of order, overview and correct notation. In many areas of engineering, a consensus has been reached to "speak the same language". As an amateur electrical engineer, this is less important, but in an international team, it is more so. Because, at the end of the day, when you need help, and someone uninvolved has to understand the thought processes, correct notation is indispensable for facilitating comprehension. That is why we are dealing with this topic here.

3.1 Notation, upper case letters, lower case letters

The most important notation rules are:

1. If an index is set, it should be meaningful.

   The car is travelling at a speed of $v_{car} = 10 \, km/h$.

2. If there are several identical sizes within an area, one distinguishes by indices. The simplest method is to number the sizes consecutively.

   Car 1 drives along $v_1 = 10 \frac{km}{h}$, car 2 drives along $v_2 = 20 \frac{km}{h}$.

3. There is no rule on how to assign indices. However, it has become accepted that an initial value is given the index zero and then numbered consecutively.

   The car drives constantly with an initial speed of $V_0 = 10 \frac{km}{h}$, then it accelerates with $5 \frac{m}{s^2}$.

4. If a variable is time-dependent, we use lower case letters. In addition, the variable on which the size is dependent is indicated in round brackets.

   The speed v of the car over time t is described by $v(t)$

For digital content, such as this book, the convention is that there is a space between the number and the unit.

> ⚠ The exception is the degree sign when we speak of an angle, but not when we speak of temperatures.

20 °C, but an angle of 180°.

5. For physical quantities, the internationally common formula symbols are used.

$$U = R \times I$$

Next, we look at how we can represent very large numbers and very small numbers.

3.2 Prefixes for a wide dynamic range

Physics uses mathematics as a tool to put events into numbers and to be able to calculate with them. Since the world covers a very large range of values, prefixes were introduced. Instead of 1000 metres, one writes 1 km, instead of 0.001 metres, one writes 1 mm and so on. The following table shows an overview of the prefixes.

Designation	Decimal number	Power notation	Name	Abbreviation
One quadrillionth	0.000000000000001	10^{-15}	Femto	F
One trillionth	0.000000000001	10^{-12}	Piko	P
one billionth	0.000000001	10^{-9}	Nano	N
One millionth	0.000001	10^{-6}	Micro	µ
One thousandth	0.001	10^{-3}	Milli	M
One	1	10^{0}	-	-
One thousand	1,000	10^{3}	Kilo	K
One million	1,000,000	10^{6}	Mega	M

Physics basics

One billion	1,000,000,000	10^9	Giga	G
One trillion	1,000,000,000,000	10^{12}	Terra	T
One quadrillion	1,000,000,000,000,000	10^{15}	Peta	P

We recall the mathematical basics. We can always write prefixes as powers and then apply the power laws.

Let's take an example in which we calculate three km (kilometres) times five mm (millimetres). First, we write both values as a superscript.

$3 \text{ km} = 3.000 \text{ m} = 3 \times 10^3 \text{ m}$

$5 \text{ mm} = 0.005 \text{ m} = 5 \times 10^{-3} \text{ m}$

$3 \text{ km} \times 5 \text{ mm} = 3 \times 10^3 \text{ m} \times 5 \times 10^{-3} \text{ m}$

The base 10 is the same, therefore the exponents can be offset. The numbers before the exponents are calculated separately.

$3 \times 10^3 \text{ m} \times 5 \times 10^{-3} \text{ m} = 3 \times 5 \times 10^3 \times 10^{-3} \text{m} \times \text{m}$

$= 15 \times 10^{3-3} \text{ m}^2 = 15 \text{ m}^2$

When invoicing, we split the numbers and their prefixes and charge them separately.

Calculate and simplify:

Three million times one billionth

Seven trillion times four thousandths

With units:

Five kilometres by eight micrometres

One teranewton times seven picometres

Solutions

$3 \times 10^6 \times 1 \times 10^{-9} = 3 \times 10^{-3} = 3 \text{ thousandths}$

$7 \times 10^{12} \times 4 \times 10^{-3} = 28 \times 10^9 = 28 \text{ billions}$

$5 \times 10^3 \text{ m} \times 8 \times 10^{-6} \text{ m} = 40 \times 10^{-3} \text{ m}^2 = 40 \text{ mm}^2$

$1 \times 10^{12} \text{ N} \times 7 \times 10^{-12} \text{ m} = 7 \times 10^0 \text{ N} \times \text{m} = 7 \text{ Nm}$

Newton is the unit of force, which we will discuss later.

3.3 The "Système international d'unités"

We have already covered the conventions of basic physics. Not only is the correct notation enormously important, but also the units with which we calculate, as the following example shows: In 1999, the Mars probe "**Climate Orbiter**" was lost when it entered the Mars atmosphere. At first, the engineers puzzled over what had gone wrong.

The solution was not long in coming and resembled a sad comedy. A NASA supplier had used the **imperial** system of units and calculated the necessary distances to land on Mars in **inches and feet**. A second NASA control team had taken the values but calculated them in **metres and centimetres**. The data were accordingly faulty, and the probe burned up in the atmosphere during the approach to Mars. This expensive example shows how important it is to use a **uniform system.** To be able to calculate physical quantities in a meaningful way, an internationally valid system of units must therefore be introduced.

In technology, it is the "**Système international d'unités**".

 In the "Système international d'unités", exactly seven **basic units** were defined. The units of quantities are therefore also called **SI units**.

The SI units were almost all defined by natural constants. Each base unit is defined by a base quantity, a formula symbol and a unit or unit symbol. The following table shows all seven base units.

Basic size	Formula symbol	Unit	Unit symbol
Time	t	Second	s
Length	s/l	Metre	m
Mass	m	Kilogram	kg
Current	I	Amps	A
Temperature	T	Kelvin	K
Amount of substance	n	Mol	mol
Light intensity	Iv	Candela	cd

The kilogram is a special SI unit. The mass of an object is given in kilograms. Contrary to what one might expect, the gram is not the base unit but the kilogram, i.e., 1,000 grams. At first, it is confusing to convert the prefixes of every other

Physics basics

unit except the kilogram. The kilogram was not defined by natural constants until 2019 but by the French "**original kilogram**".

It should also be noted that there are "naturalised" units. For example, one thousand kilograms is called one ton = 1,000 kg = 1 t.

For units of length and area, the prefixes for centimetre (1 cm = 0.01 m), for decimetre (1 dm = 0.1 m) are also often used.

3.4 Derived SI units

In physics, there are many other quantities, such as the area A, the force N or the voltage U. All these quantities can be derived from the base SI quantities. Therefore, they are called **derived SI units**.

The area A is a derived SI-unit

$$\text{Area} = \text{Length} \times \text{Width with m} \times \text{m} = \text{m}^2.$$

The force N with the unit Newton is derived from

$$Force = \frac{Mass \times Length}{Time \times Time} \text{ with } N = \frac{m \times l}{s^2}$$

It is common to write a physical quantity in square brackets and then state the unit.

For the purpose of this book, this convention will be maintained; if something is in brackets, it is a unit.

For example, the unit of time is the second $[t] = s$.

3.5 Displaying differences

If one wants to represent the difference of a quantity in physics, a **delta Δ** is used. For the difference between two quantities of energy, for example, one writes $E_2 - E_1 = \Delta E$.

The large delta describes a **difference.**

The differential

If we let this delta become smaller and smaller in our thoughts, the values E_2 and E_1 become closer but **never** become exactly the same. For this approximation of an infinitely small difference, we use a **differential**. The large delta becomes a small d.

$\Delta E \rightarrow dE$

The change from one size to another is written as a differential. For example, speed is equal to the change of distance after time. As a differential representation:

$$v = \frac{s_2 - s_1}{t_2 - t_1} = \frac{\Delta s}{\Delta t} \rightarrow \frac{ds}{dt}$$

3.6 Conservation of energy and efficiency

Most physical laws can be considered "natural". You can extrapolate these laws until you reach the level of the smallest particles. As always, however, within this framework, the emphasis is not on detailed derivation but on understanding and practical examples. This is also the case with the **conservation of energy**.

Energy conservation means that there is a fixed amount of energy in the universe and that this energy cannot be destroyed or generated. Energy can only be converted into different forms. We already know most forms, such as heat energy or kinetic energy.

For example, in a wind power station, the kinetic energy of the wind is absorbed as rotational energy and then converted into electrical energy.

One limitation is that each conversion produces a proportion of unusable energy, usually in the form of heat. The ratio of how much energy is retained during conversion and how much is lost describes **the efficiency**. The efficiency is a dimensionless quantity and is indicated by the Greek letter η (eta). η is defined as the ratio of the usable

$$\eta = \frac{E_{usable}}{E_{total}}$$

When converting energy from one form to another, the efficiency describes the ratio of the usable energy after the conversion to the total energy before the conversion.

$$\eta = \frac{E_{after}}{E_{bevore}}$$

It is easy to see that η is always between zero and one. Often, η is given as a percentage. At an efficiency of exactly 1 (= 100%), all energy is converted without loss. At an efficiency of zero, the entire energy is no longer usable after conversion.

In connection with efficiency, the terms *exergy* and *anergy* are also often used. Exergy describes the part of energy that can be used. When driving a car, this is the part of the energy that is converted into propulsion for locomotion; the waste heat, i.e., the heating of the engine, the unused energy, is called anergy.

Examples from everyday life:

Photovoltaic modules today have an efficiency of about 20%. This means that only one fifth of the solar energy is converted into electricity. And this electricity must then, in turn, be converted to the appropriate voltage for the socket or stored, whereby losses in the range of 2-10% can occur.

A classic combustion engine has an efficiency of around 45%, whereas an electric motor, such as those used by VW, Tesla or Audi, has an efficiency of more than 90%!

LEDs convert about 40 - 50% of the electrical power into light. The rest is needed by the control electronics to stabilise the current flow and is converted into heat. With conventional incandescent light bulbs, the efficiency is significantly lower. Here, only 10 - 20% of the energy is converted into light, the rest is released as heat.

However, a coal-fired power plant, which is supposed to convert the energy from the coal solely into electrical energy, generates over 60% thermal energy, i.e., an efficiency of just under 40%. If several processes or conversions are carried out in succession, the total efficiency is obtained by multiplying the individual efficiencies.

Example: A wind turbine can convert 50% of the kinetic energy of the wind into rotational energy. The speed of the rotor is then increased by means of a gearbox. The gearbox has an efficiency of 95%. The generator, which finally provides the electrical energy, has an efficiency of 90%. The overall efficiency of the wind turbine is thus as follows

$\eta_{total} = \eta_1 \times \eta_2 \times \eta_3 = 0.5 \times 0.95 \times 0.9 = 0.4275$ (corresponds to 42.75%)

Efficiency is an important aspect when it comes to developing and promoting technologies, as it is often closely related to economic viability.

However, the context must always be taken into account. Solar and wind energy are available in unlimited quantities and free of charge. Therefore, an efficiency of 20% or 40% is not a knock-out criterion for this technology. Gas or petrol, on the other hand, are raw materials and, therefore, have a market price, which is why you have to extract as much exergy as possible from the raw materials in order to be able to operate profitably.

If we burn oil worth €100, at least €100 worth of electricity must be generated in the process. Every percent increase in efficiency goes directly into the economic balance. Another faux pas that is very often committed by technically unskilled people is the confusion of power and energy.

Especially when it comes to renewable energies, these terms are often used incorrectly, as are inappropriate quantities and units, so that engineers can only grab their heads in incomprehension. Therefore, the difference is clearly explained below.

3.7 Energy

The terms energy and work are used in physics for the same physical quantity. The energy, or the work done, is abbreviated as E or W.

 Work describes the process of converting one form of energy into another. For example, work is done when a heavy stone is lifted. The energy given refers to the stored work within a system: the stone has potential energy after being lifted. Practically and mathematically, the terms are used in the same way. The same units and formulae are used for calculation.

Energy can be thought of as the stored water in a tank: It is a fixed quantity. Its SI unit is the joule [J]. $1 J = 1 \frac{kg \times m^2}{s^2}$.

Alternative units are watt-seconds Ws or kilo-watt-hours kWh. Naturalised units of energy are, for example, kilo-calories kcal. With kilocalories, we indicate the energy in our food.

3.8 Power

Power, on the other hand, is a physical quantity and refers to the energy or work that is converted or performed in a certain amount of time.

$$P = \frac{\Delta E}{\Delta t}$$

The unit of power is the watt, corresponding to energy per time. The unit of energy is the joule, the unit of time is the second.

One watt is therefore equivalent to one joule per second. $1 \frac{J}{s} = 1 W$.

On our electricity bill, the energy is stated in kWh, i.e., the amount of kW of power used over a certain number of hours.

Excursus: Horsepower

Another unit of power is horsepower.

The unit horsepower goes back to James Watt. Horsepower described the average continuous performance of a working horse. It is unclear which horse and which power measurement was chosen as a reference. There are many assumptions, one being that James Watt used a pit horse as a yardstick. The horse pulled coal sacks out of the pits via ropes and pulleys. The working time, the weight of the coal sacks and the height lifted were used for calculation. The result was that one hp corresponded to approximately 735 W.

Ultimately, the unit of horsepower only became established in manufacturing engineering. In physics, the watt, named after James Watt, is used almost without exception.

If power P is effective over a period t, energy E is converted:

E = P x t.

The correct formulation in Physics is:

W = P x t (the work done).

However, in the context of this book, we use the common formulation of energy and dispense with this formulation to facilitate understanding.

Example: A hairdryer has a power consumption of 2000 W or 2 kW. If you let the hairdryer run for one second, the hairdryer consumes (actually not correct, because the energy is converted into moving heat and not "consumed") an energy of

$2000\ W \times 1\ s = 2000\ Ws = 2\ kJ$.

After half an hour the hairdryer has used up 2 kW x 0.5 h = 1 kWh, after one hour it is 2 kWh and so on. The power remains constant the whole time at 2 kW while the energy depends on the time elapsed.

If you want to convert J or kJ into kWh or vice versa, the following conversion applies:

$1\ J = 1\ Ws = 1\ W \times \dfrac{1h}{3600\ s} = 2.8 \times 10^{-7}\ kWh = 0.00000028\ kWh$

$1\ kWh = 3600\ kWs = 3600\ kJ$

Test yourself: Which units are correct, which are wrong? It's not about whether the numbers are right, but only about the units!

In one hour, a refrigerator consumes 100 W.
Wrong - A refrigerator has a power consumption of 100 W. In one hour, it consumes 100 W. Accordingly x 1 h =0.1 kWh

The maximum demand for electrical power (in Germany) is approximately 80 GWh.
Wrong - The unit for power is W. So, the correct figure is 80 GW.

A Tesla Model 3 has an engine power output of 360 kW and a battery capacity of 75 kWh.

Both are correct – the unit for power is W (360 kW corresponds to approx. 490 hp), the energy stored by the battery in kWh. (The term battery capacity is not physically correct, as capacity is a different measure. Colloquially, it refers to the amount of energy stored).

100 g of bread contains an energy of about 1 MJ.

Correct - even though the unit joule is unusual for food, it is a form of energy. One MJ corresponds to about 240 kilocalories.

Herbert consumes 150 W when cycling. How much energy does he use in two hours of cycling? How many kcal is that? (1 kWh = 860 kcal)

The energy results from the power multiplied by the effective time. Therefore, a total of 300 Wh = 0.3 kWh. This corresponds to 258 kcal. However, due to losses during conversion, Herbert burns considerably more than 258 kcal in practice.

A trained young man can produce about 100 W of continuous power. We recall that a pit horse can produce one horsepower, which corresponds to about 735 W.

4 From water model to circuit

If you are new to the subject of electrical engineering, many of the terms used are abstract and difficult to imagine. It takes time and practice to define the terms and classify them correctly. To make it easier to memorise the terms, we use a model for this.

 A model is a **simplification of reality** and attempts to map new, complex facts to what is already known.

In our case, one can transfer the topic of the electrical circuit to a known **water circuit** by many analogies. Each component in the water circuit is compared with a corresponding component from the electrical circuit:

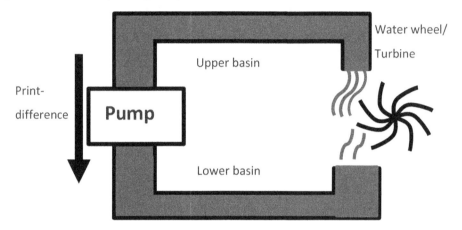

Figure 8 Water cycle

A water circuit consists simply of two water basins, a water pump, pipes that transport the water, and a consumer, for example a turbine, a water wheel or something similar.

One water basin is **situated higher** than the other. The pump is constantly pumping water upwards. As a result, the water in the upper basin has greater potential energy. There is a pressure difference between the upper and lower basin.

The water runs **through the pipes** and via the consumer back into the lower basin. The consumer is driven by the moving water. The water thus transfers the **energy of the pump** to the consumer. For each element in the water circuit, we look for a corresponding element in the electrical circuit.

Let's start with the pipes through which the water flows. In the electrical circuit, these are the **cables or pipes** through which the current flows. The water in the water circuit corresponds to the current, which consists of **moving electrons.** But how is a conductor actually constructed, and how can the electrons move in it? To find out, let's take a very close look at the material.

4.1 Atoms, electrons, protons

To understand the different effects of electrical engineering, we should first take a look at the basic building blocks of physics – atoms. Every material consists of atoms at the smallest level. An atom consists of positively charged particles, **protons**; particles without charge, **neutrons;** and negatively charged particles, **electrons.**

The symbol of the **charge is Q** and the unit of the charge is the **coulomb C**. In SI units $1\ C = 1\ A \times 1\ s =$ As. The charge is therefore indicated in C or As.

The **elementary particles** (protons and electrons) both have the smallest charge that is physically possible. This is called the **elementary charge** and is abbreviated with e – not to be confused with Euler's number, which is also abbreviated as **e.** This is confusing, but you can usually tell from the context what the abbreviations are.

The elementary charge has the value of **e = 1.602 x 10^{-19}** Coulomb. An electron has a charge of Q = **-e** and a proton has a charge of Q = **+e**.

Since the charge of an atom is neutral overall, it has **the same number of electrons as protons**. The protons and neutrons form the nucleus of the atom, while the electrons race around the nucleus at the speed of light. Almost the entire mass of the atom is united in the atomic nucleus.

Each element, such as hydrogen, oxygen, carbon and iron, or even nickel, copper and zinc, has a very specific, unique number of protons and electrons that hold the element together.

The smallest and lightest element is hydrogen. It has the **atomic number one**, which means that it consists only of a single proton and one electron. It has no neutrons. Iron, on the other hand, has the **atomic number 26**, which means that it consists of 26 protons. There are also 30 neutrons in the nucleus so that the iron atom is about 56 times as massive as the nucleus of a hydrogen atom.

A molecule such as carbon dioxide, for example, is formed when several individual atoms combine, in this case one carbon atom with two oxygen molecules to form CO_2, hence the name carbon di (two) oxide (oxygen compound). Molecules can be formed by various reactions even under normal conditions. For example, rusting of metal is a reaction of iron Fe and oxygen O to form iron oxide Fe2O3 – the red rust layer. This basic mode of operation, that elements combine with other elements through external influences and thereby absorb or release energy, is essential for our life and will become even more important.

If you go into more detail about the structure of an atom, you can see that the electrons do not race randomly around the nucleus of the atom but are on defined orbits.

Figure 9 Structure of an atomic model

These are known as **shells**, which can hold different numbers of electrons. The innermost shell (K-shell), which is close to the nucleus, can hold only two electrons; the second (L-shell) can hold up to **eight electrons**, the third (M-shell) up to **18 electrons** and so on. In total, there are up to **seven shells**, depending on how many protons and thus how many electrons an atom has.

If an atom has only two electrons, only the first shell is filled. If it has 11 electrons, the first and second shells are completely filled, and in the third shell there is a single electron.

4.2   When does material conduct electricity?

Electricity consists of nothing but moving charge carriers. A material is therefore highly conductive if the charge carriers can move easily. Since the protons are fixed in the nucleus, only the electrons remain, which can move freely, but they are attracted by the positive nucleus.

Since the electrons on the outer shells are not so strongly attracted, they can more easily separate from the atomic nucleus.

 The electrons on the outer shells are therefore of great importance for the conductivity of a substance.

The electrons on the outermost shell are also called **valence electrons**.

Metals such as iron, copper or aluminium form a special grid structure in which the valence electrons can **move freely**.

In metals, the valence electrons buzz around in the lattice like a homogeneous gas; one also speaks of an **electron cloud** or **electron gas** in the metal.

Non-conductive materials, such as most plastics, do not form a lattice and retain their valence electrons to a large extent. This prevents electrons from flowing through the material.

What is generally known as current is nothing other than the movement of the valence electrons from A to B.

 A current flow consists of moving charge carriers.

Let's get back to our water model. The valence electrons are freely movable and therefore correspond to the water in the water cycle. They transfer charges or energy in the cycle. Next, we come to the **pressure difference** between the pools. This is caused by gravity — mathematically speaking, by the **gravitational field of the earth**. Analogous to this is the **electric field** in a circuit.

5 The electric field

First, let's clarify the properties of the Earth's gravitational field. This ensures that everything on this planet experiences an attraction towards the centre of the Earth. The principle behind this is that masses attract each other. The larger the masses and the closer the masses are to each other, the stronger the force of attraction.

In our water cycle, this means that the water can drive the turbine because it was pumped up by the pump, i.e., it was lifted against gravity or the earth's gravitational field. Physically speaking, work has been done and potential energy has been supplied to the water. This creates a pressure difference. When the water flows down the pipes, the pressure is converted at the consumer (the turbine).

Analogously, in electrical engineering, there are electric and magnetic fields that assign potentials to electrons. But what is a field, and how can it be imagined?

5.1 Representation of E-fields

First of all, every field has a cause.

 In the electric field or simply E-field, the cause is charged particles. Electric fields form around charged particles.

An accumulation of positive charge is called a positive pole, an accumulation of negative charge carriers correspondingly a negative pole.

In order to be able to represent the field, draw field lines that start at the cause.

 Field lines always point away from a positive charge and towards a negative charge. The density of the field lines indicates the strength of the field.

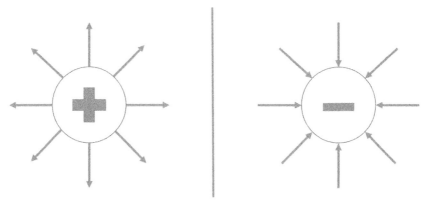

Figure 9 Field lines of a point charge

The illustration in Figure 10 shows that the field lines (light blue arrows) are clearly closer together (denser) at the circle representing the point charge than away from it. This means that the field is correspondingly stronger there.

If several charge carriers meet, a wide variety of field lines are created (Figure 11).

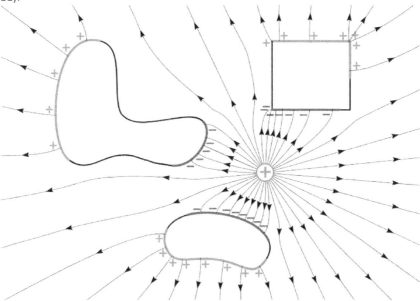

Figure 10 Electric field lines

The field lines are jumbled and do not seem to follow any order.

If, on the other hand, the field lines are parallel, we speak of a homogeneous field. The field has the same value at every point. This is the case, for example, if we have two opposing metal plates on which charge carriers are placed (Figure 12).

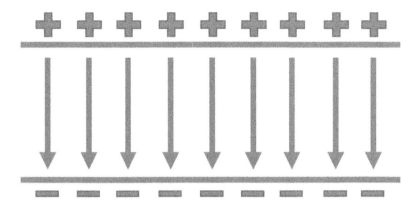

Figure 11 Homogeneous electric field

5.2 The force in the electric field

If we place a sample charge, for example a proton, in the field, it is repelled by the positive pole and attracted by the negative pole. The particle thus experiences a force along the field lines. Based on this fact, we can define the electric field strength E which is the force F exerted by the field on a sample charge Q.

$$E = \frac{F}{Q}$$

The field is also often written as a vector. $\vec{E}$

This is because the field not only exerts a force on the particle, but the force also has a direction in space.

 The calculation always refers to the electric field strength. In colloquial language, only "electric field" is used. Strictly speaking, this is not correct, as the "field" only describes the spatial distribution but not its strength.

The unit of the electric field strength is therefore:

$$[E] = \frac{N}{C} = \frac{kg}{m^2 \times As}$$

Another unit for the strength of the electric field is volts per metre $\frac{V}{m}$.

Summary of an Electric field:

An electric field is formed wherever electric charges are present.

To illustrate this, draw field lines that run away from positive charges and towards negative charges. The density of the field lines corresponds to the strength of the field.

The electric field exerts a force on a sample charge.

What force is experienced by a single proton with a charge of
$Q = 1.602 \times 10^{-19}$ C in an E-field with $E = 3,000,000,000 \frac{N}{C}$?

Solution:

$E = \frac{F}{Q}; F = E \times Q = 3 \times 10^9 \frac{V}{C} \times 1.602 \times 10^{-19} C = 4.8 \times 10^{-10} N$

$= 480 \text{ pN(Pikonewton)}$

What force does an electron experience in the same E-field? What is the difference?

Solution: A proton has the same charge as an electron but a different sign. Therefore, the force is the same for the electron but with a negative sign (-480 pN). The proton is accelerated in the opposite direction.

The electric field

Excursus equipotential lines:

In more in-depth literature, equipotential lines are also often mentioned. These are perpendicular to the electric field lines (Figure 13).

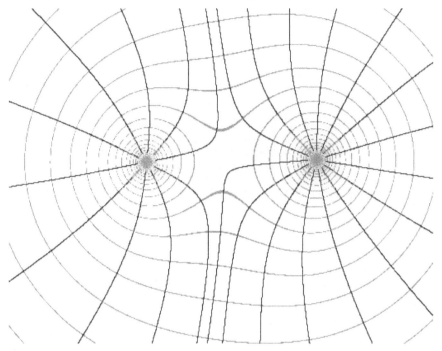

Figure 12 Equipotential lines

The dark lines are the field lines of the point charges. The ellipses represent the equipotential lines. At the **crossing points,** equipotential and E-field lines are **perpendicular to each other**. The same electrical potential exists at each point. To understand the meaning of the equipotential lines, we must first learn about electric potential and voltage U.

5.3 Electric potential and voltage

Electric potential, also called electrostatic potential, is abbreviated with φ (Greek lower case letter Phi). It has the unit volt V.

 Electric potential describes the potential energy of a sample charge within an electric field. The electric field assigns a potential to every point in space.

Analogous to the water cycle, it is the absolute pressure that the water possesses and exerts through height. The pressure exerted by the basin of water at a certain height is determined by the earth's gravitational field. As a simplified example, the upper basin has a gravity pressure of one bar and the lower basin has a pressure of zero bar.

However, in the water circuit, it is not the absolute pressure that matter, but only the relative pressure, i.e., the pressure difference between the lower and the upper water basin.

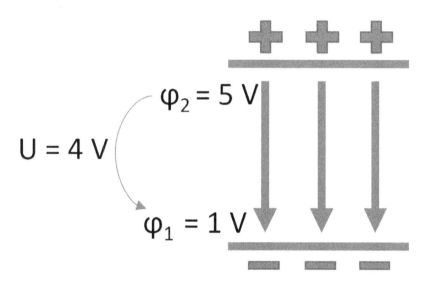

Figure 13 Potentials and voltage in the homogeneous E-field

The difference between two potentials $\varphi_2 - \varphi_1$ is called the voltage U. Voltage also has the unit volt V.

! Voltage only indicates a potential difference. Therefore, you always need a reference potential.

But what exactly is our pump now? The pump in the water circuit corresponds to the voltage source in the electrical circuit. A voltage source is, for example, a battery. A standard AA battery has a voltage of 1.5 V. This means that the positive pole, i.e., the upper contact point of the battery, has an electrical potential that is 1.5 V higher than the negative pole.

The circuit symbol of a voltage source is a circle with a solid line. Every voltage source consists of a positive and a negative pole. An ideal voltage source generates a voltage independent of the applied load. In reality, this is only approximately possible.

In electrical circuits, one usually chooses the lowest potential and specifies it as the reference potential. This means that it has the potential of ϕ = 0 V, and all other potentials are specified in relation to this potential.

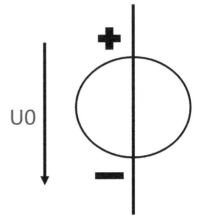

Figure 14Switching symbol of a voltage source

 In electrical engineering, voltages are relevant almost without exception. Potentials are hardly ever considered, as current can only flow at a potential difference

5.4 Current

We have already learned that the water in the water cycle corresponds to electrons in an electrical circuit. But in everyday life we always talk about current, i.e., moving electrons. A measure of the strength of the electron flow is therefore the current strength, abbreviated by the formula symbol I. Its unit is the ampere A. Since the current indicates the flow of electrons, and each electron has a charge, the current indicates how much charge is transferred per time.

$I = \frac{Q}{t}$ the unit ampere is accordingly $1 \, A = 1 \frac{C}{s}$

 What is the electric current when one quadrillion (10^{15}) electrons each with a charge of $e = 1{,}602 \times 10^{-19}\, C$ flow?

Solution: First we calculate the charge. The total charge is the charge of an electron multiplied by the number of electrons.

$$Q = n \times e = 10^{15} \times 1.602 \times 10^{-19}\, C = 1.602 \times 10^{-4}\, C = 160.2\, \mu C$$

Then we look at the time in which this charge flowed.

$I = \frac{Q}{t} = 160.2 \frac{\mu C}{1s} = 160.2\, \mu A$

Power sources

Analogous to voltage sources, there are also current sources. These do not produce a potential difference but a constant current I, independent of the voltage U applied. In the water circuit, we can imagine a current source as a pump that always generates a constant amount of water flow, i.e., it only drives the water but does not raise it or increase the pressure.

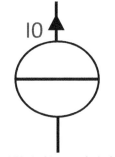

Figure 15 Switching symbol of a current source

5.5 Technical and physical current direction

In electrical engineering, as well as all other engineering sciences, the term "technical direction of current" is used, but what does this mean exactly?

Let's start with the physical direction of the current. We know that electrons are negative charge carriers that form the current flow. Therefore, the current flows from where there are more electrons to where there are fewer electrons. Since electrons are negatively charged, the current flows from the negative pole to the positive pole. This is the "physical" direction of current or the direction of the flow of the electrons.

 The term "physical direction of current" is somewhat misleading since in physics the technical direction of current is usually used. The "physical direction of current" merely corresponds to the direction of movement of the electrons, not the direction actually used in electrical engineering.

However, the current and its properties were discovered before it was known exactly whether the positive or negative charge carriers were responsible for the current flow. It was wrongly assumed that the positive charge carriers, i.e., the protons, form the current flow. In the model conception, the current, therefore, flowed from the positive to the negative pole. This notion has been retained until today. Nothing changes in the calculations. It is only good to know that the current flow in reality is different from the way we draw it.

 The technical current direction is used in all circuit diagrams, drawings and circuits.

In order not to get completely confused, let's remember:

 In a technical circuit, the current always flows from the positive pole to the negative pole!

The electric field

6 The magnetic field

Just like the gravitational field, the magnetic field is a familiar phenomenon in our everyday lives. Everyone knows magnets, for example when they are used to attach notes to a pinboard. Since these magnets are permanently magnetic, they are also called permanent magnets.

There are many similarities and analogies between magnetic and electric fields. An overview of the comparison between magnetic and electric fields can be found at the end of this chapter.

The magnetic field strength has the formula symbol H. Since it also has a direction, just like the electric field, it is often described as $\vec{H}$. The unit of the magnetic field is $\frac{A}{m}$.

Often, it is not the absolute magnetic field that is needed but the magnetic flux density $\vec{B}$. This indicates how strong the magnetic flux in the magnetic field is. It also indicates the force acting on a sample charge.

We are not interested in the complete magnetic field of a body but only in the "effects", and this is described by the flux density.

You can imagine the magnetic field like a waterfall. We are not interested in the complete extent and size of the waterfall but only in the flux density of the falling water.

The magnetic field strength $\vec{H}$ is less important in technology. Almost without exception, flux density $\vec{B}$ is calculated.

Therefore, one generally speaks of a B-field as an abbreviation for the magnetic field (analogous to the E-field - the electric field).

The unit of magnetic flux density is the Tesla T, or Newton per ampere and per metre. $1\,T = 1\,\frac{N}{A \times m}$

The magnetic flux density and the magnetic field are directly related via the **permeability μ.** μ is therefore also often called magnetic conductivity.

$$\vec{B} = \mu_0 \mu_r \times \vec{H}$$

μ_0 = Permeabilty in vacuum = $1.257 \times 10^{-6} \frac{Vs}{Am}$

μ_r = Substance – dependent permeability

Common permeabilities μ_r are, for example, iron or ferrite with μ_r to 15,000.

Now that we have learned about the physical quantities, we come to the cause of a B-field. An electric field is created when charged particles form a positive and a negative pole.

 The cause of a magnetic field in a permanent magnet is not charged particles but so-called elementary magnets.

6.1 Elementary magnets

This is again a physical model. Each element consists of countless small elementary magnets. These elementary magnets cannot be broken apart because they represent a small unit. Just like a "large" magnet, they consist of a north and a south pole. Equal poles repel each other; different poles attract each other.

In most materials, these elementary magnets are arranged without a system. The respective poles neutralise each other, and the material is not magnetic.

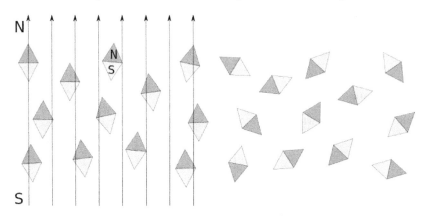

Figure 16 Elementary magnets

In magnetic materials, all the elementary magnets are aligned. This creates a north and a south pole: the material is magnetic. The best-known magnets are neodymium magnets. These are made of the element neodymium (Nd), which belongs to the rare earths, iron and boron. Due to their extreme strength, neodymium magnets are used in many areas, for example in the asynchronous generators of wind turbines or in the drives of electric cars.

Excursus: Magnetising materials

You may know that you can magnetise certain non-magnetic metals with the help of a permanent magnet. If you rub the metal several times with a permanent magnet, it gradually becomes slightly magnetic. In the process, the permanent magnet aligns the elementary magnets in the metal in one direction. Over time, the elementary magnets re-arrange themselves and remain in this position. North and south poles are created, and the metal is magnetised.

6.2 Displaying magnetic fields

Just as with the electric field, the magnetic field is represented by field lines.

 Unlike electric field lines, magnetic field lines are always self-contained, i.e., they do not have a starting and end point.

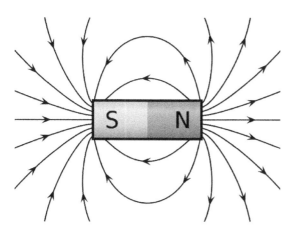

Figure 17Magnetic field lines of a permanent magnet

The illustration in Figure 18 shows the magnetic lines of a permanent magnet. But these are not self-contained, are they?

Yes, they are closed because the magnetic field lines continue inside the magnet from the south to the north pole so that a closed circle is formed. We are only interested in the outer field lines, which is why many illustrations only show the outer magnetic field lines.

We can draw the magnetic field lines as an arrow from one pole to the other, knowing that the field lines continue inside the magnet. For the outer magnetic field lines, it is then true that the starting point is always the north pole and the end point the south pole.

The density of the magnetic field lines indicates, as in any field, the strength of the magnetic field and is, therefore, a measure of the flux density $\vec{B}$.

A B-field is distributed three-dimensionally in space. When drawing on a two-dimensional plane, for example on a sheet of paper, a convention has become established. A magnetic field that points into the drawing plane is indicated by a cross. If it points out of the drawing plane, it is marked by a dot.

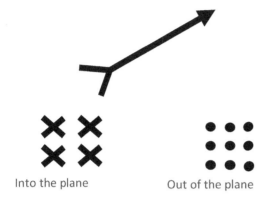

Into the plane   Out of the plane

Figure 18Representation of field lines

You can remember this definition by thinking of an arrow. If you shoot the arrow into the plane, you see the tail, which is a two-dimensional cross. If the arrow flies towards us, we only see the tip, i.e., a point.

6.3 Electromagnetism

Static magnetic fields, such as a permanent magnet, are easy to explain and well-known. Magnetic fields generated by electricity, for example those in an electric motor, are much more complex and difficult to understand.

Electromagnetism is one of the most important phenomena of our times and can be found almost everywhere. Electromagnetism plays a role in the electric car, in data transmission, in the high-voltage transmission of our power grid and in every power supply unit of a PC, laptop or smartphone.

To understand this, let's go back in time a little. In 1820, the physicist Hans Christian Ørsted experimented with a piece of wire through which he let current flow. In the process, he noticed that a compass located nearby deflected each time the voltage was applied. The magnetic needle no longer pointed north but was deflected by the wire through which current was flowing. This discovery quickly made the rounds and other physicists, such as André-Marie Ampéré, after whom the current bar was named, were able to confirm the experiment.

This proved that an electric current generates a magnetic field.

A current-carrying conductor generates a magnetic field.

But how do the field lines of the magnetic field run?

After some investigation, it was found that the resulting B-field builds up concentrically around the conductor through which the current flows.

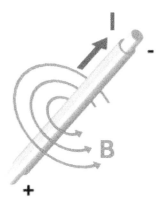

Figure 19Magnetic field of a current-carrying conductor

The direction of the B-field can be determined with the right-hand rule. This involves clenching a fist with the right hand and pointing the thumb upwards. This indicates the direction of flow of the current (the technical direction of the current, i.e., from positive to negative pole). The four fingers indicate the direction of circulation of the B-field.

6.4 Induction law

In physics, most effects are valid in both directions. A magnetic field is generated by a current flow in the conductor.

Conversely, an externally applied magnetic field generates a current flow in a conductor. This inversion is described as the law of induction. The process of electromagnetic induction means that an external, changing magnetic field generates a current or voltage in a conductor.

Derivation of induction

 To do this, we learn a new quantity: We already know the magnetic field and the magnetic flux density. The third significant quantity is the magnetic flux Φ (Greek letter capital Phi).

The magnetic flux can be compared to a waterfall. We take a surface and hold it in the waterfall. We look at how much water flows through the surface.

The quantity that flows through the area corresponds to the magnetic flux. We obtain the magnetic flux by multiplying the flux density by the area flowed through.

$$\Phi = \vec{B} \times \vec{A}$$

 This relationship only applies to homogeneous magnetic fields, but in the context of this book we will limit ourselves to this "special case".

The unit of magnetic flux is described by Tesla times square metre Tm^2 or Weber Wb ($1\ \mathbf{Wb} = 1\ \mathbf{Tm^2}$).

 We have a homogeneous magnetic field with a flux density of B = 200 mT. If we consider a square surface with an edge length of 10 cm. How large is the magnetic flux?

Solution: $0.2\ T \times 0.1\ cm \times 0.1\ cm = 2\ mWb$

6.5 Magnetic flux and induction

The law of induction states that the voltage induced on a conductor depends on the change in magnetic flux over time. The temporal change is described by the differential

$$U_{ind} = -\frac{d(B \times A)}{dt}$$

In practical terms, this means that a voltage is induced on a conductor when:

1. the magnetic flux B changes over time. This can be the case, for example, when more energy is supplied to an electromagnet, and the field becomes larger as a result,

2. the area A through which the magnetic field passes changes. This can happen, for example, when the surface is pulled out of the B field or immersed.

The second effect is used, for example, in the dynamo of lights for bicycles: A permanent magnet rotates past a conductor. This causes the conductor to dip in and out of the B-field with each rotation. According to the law of induction, a voltage is induced that operates the front and rear lights.

 With the help of this system, a voltage can therefore also be induced from a movement. The law of Induction is the basis for electro-mechanical systems such as electric motors and generators.

6.6 The Lenz Rule

 Nature is "lazy" and does not like to change. It strives for balance and homogeneity. This can be observed in many natural effects.

In electrical engineering, there is also an effect described by Lenz's rule. It states that an induced voltage counteracts its cause (the change in the B-field or the area).

This can be explained by the following example.

A conductor is completely in a B-field.

> Since the B-field does not change and the conductor lies completely in the B-field, no voltage is induced. Let us now assume that the external magnetic field decreases due to external influences.

- The change in the magnetic field is no longer zero.

- A voltage is induced in the conductor.

- A current flows through the induced voltage.

- This generates a magnetic field that is superimposed on the external magnetic field.

- However, the magnetic field generated is polarised in such a way that it counteracts the cause, i.e., the decrease of the external B-field. Accordingly, it has a "building up instead of a decreasing" effect.

- The generated magnetic field "supports" the external magnetic field so that it decreases more slowly.

- As a consequence of Lenz's rule, no abrupt changes of the magnetic flux are possible.

The increase and decrease of the magnetic flux induce a voltage that counteracts the cause.

6.7 The Lorentz force

In an electric field, a sample charge Q experiences a force that pulls the sample charge towards one pole and repels it from the other.

As in an electric field, test charges that are placed in a magnetic field experience a force. A test charge is not a charge particle, but a magnet. And as we have learned, a current-carrying conductor is also a magnet because it generates a magnetic field.

 A force, the Lorentz force, acts on a current-carrying conductor in the magnetic field.

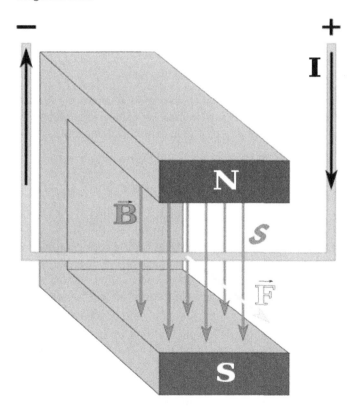

Figure 20Lorentz force in a horseshoe magnet

As an example, we will use a horseshoe magnet (see Figure 21). The advantage of this magnet is that the field within the legs of the horseshoe can be considered approximately homogeneous.

The B field flows from the north pole to the south pole. If we place a current-carrying piece of wire in the homogeneous magnetic field, it experiences a force. The magnitude of the force depends on the strength of the B-field B, the strength of the current I, and the length of the piece of wire s.

$$F_L = I \times B \times s$$

What force is experienced by a conductor with a length of 10 cm, through which 10 A flows and which is in a B field with a flux density of 200 mT?

Solution: $F_L = 10\ A \times 0.2T \times 0.1\ m = 0.2\ N$

What is the length s of a wire on which seven milli-newtons of force act when it is in a B-field of strength $100\ mT$ and a current of 200 mA flows through it?

Solution:

$$F_L = I \times B \times s$$

$$s = \frac{F_L}{I \times B} = \frac{7\ mN}{0.2\ A \times 0.1T} = 35\ cm$$

6.8 The three-finger rule

The direction of the force can be determined with the right-hand or three-finger rule. The thumb, index finger and middle finger are stretched out to form a right-handed coordinate system (right-hand system). The thumb is the direction of the technical current, the index finger is the direction of the magnetic field, and the middle finger indicates the direction of the resulting force.

The magnetic field

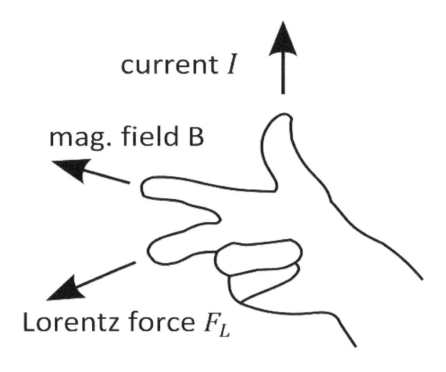

Figure 21 Three-finger rule

6.9 Overview: E-field and B-field

Finally, the following table illustrates all the analogies of electric and magnetic fields. Strictly speaking, the Lorentz force is not a purely magnetic force effect since a current-carrying conductor is necessary.

Type	E-field	B-field
Field strength	$\vec{E}$	$\vec{B}$
Field displayed graphically		
Field constant	$\varepsilon_0 = 8.854 \times 10^{-12} \dfrac{As}{Vm}$	$\mu_0 = 1.257 \times 10^{-6} \dfrac{Vs}{Am}$
Cause	Charged bodies	Permanent magnets or current-carrying conductors
Test specimen	Test loading	Sample magnet/ current-carrying conductor
Field lines	Line along which a specimen experiences a force	
Field line orientation	From positive to negative pole	Closed, outside from north to south pole
Force effect	Coulomb force $F_{el} = E \; x \; q$	Lorentz force $F_L = I \; x \; B \; x \; s$

We have now become acquainted with the most important basics. In the following chapters we will deal with the representation and modes of action of concrete, electrical components.

7 Markings and treasure symbols

We have already learned that there must always be a reference potential in electrical systems since voltages only indicate a potential difference.

7.1 Mass and earth

The "zero potential" in a circuit is also called earth or GND (ground). Every electrical circuit has a ground potential. The circuit symbols for the ground potential are as follows:

Figure 22 Symbol 02-15-04 according to DIN EN 60617-2 for mass and 02-15-04 according to DIN EN 60617-2 for mass, enclosure

Often the potential of the earth is set as the ground potential. This can be done, for example, by anchoring a stake in the ground, similar to the way lightning rods are anchored. It may sound strange at first to include the earth as an "electrical element" in a circuit, but in practice this is quite common. Especially for protective measures, a reliable (down) conductor is needed. The earthing can be found, for example, in ordinary household sockets. The metal pins at the top and bottom are attached to earth the operated device so that, for example, the housing of a connected device cannot become charged or dangerous. The circuit symbols for earthing are very similar to those for ground.

7.2 The consumer

We have already described most of the analogies between electricity and water circuits: The pipes correspond to the tubes, the pressure difference to the voltage and the pump to a voltage or current source.

A further analogy is a turbine or the water wheel that extracts energy from the water in a water circuit and something that offers resistance to the current in an electrical circuit. This can be anything: a lamp, a mobile phone while charging, a refrigerator or even a television. This is known as "the consumer", the symbol of which is a circle with an inner cross.

Figure 25: The circuit symbol of a consumer, for example, a light bulb:

Figure 23 Switch symbol consumer

7.3 The completed circuit

We have learned about all the essential components of the simplest type of circuit, and we have compared electrical and water circuits. An overview of the analogies is presented in the table below.

Water circuit	Representation	Electric circuit	Symbol
Pump		Voltage/ Power source	
Pipes		Cables	
Waterwheel Turbine		Consumer lamp, refrigerator etc.	
Water particles	H2O	Electrons	e-
Water flow	agitated water	Current flow	I
Water pressure	p	Potential	ϕ
Pressure difference *(between the basins)*	Δp	Voltage	U
Lower basin	"Zero level"	Ground/earth	

Markings and treasure symbols

Of course, there are many other components in an electrical circuit, some of which we will get to know later. For some, there are analogies to the water circuit, for others not. In the next chapter, we will look at some of these components. Before that, however, we will learn about the conventions for drawing voltage and current arrows as well as two fundamental laws in electrical engineering.

7.4 What happens without consumers?

If there is no consumer in the water circuit that slows down the water, the water is theoretically accelerated. The pump works at full throttle until the lower basin is empty or the water has absorbed so much energy that the pipes can no longer withstand the pressure, and the circuit breaks apart, for example by a pipe bursting. The same happens in an electrical circuit. Without a consumer, there is nothing to stop the current. It gets stronger and stronger until the cables get so hot that they burn up. Everyone knows this phenomenon better as a **short circuit**.

A short circuit occurs when positive and negative terminals are connected without a consumer! A simple conductive circuit is, therefore, a short circuit because there is no consumer in the circuit.

7.5 Counting arrow systems

We have already learned about voltage sources, current sources, lines and consumers, but how can we calculate with these symbols? In more complex systems, it is important to use a unit system. We remember the burn-up of a probe in the atmosphere of Mars because two different unit systems were used.

In order to be able to calculate, a uniform arrow system for voltages and currents is necessary.

7.6 Voltage arrows

Voltage arrows have a starting point and an end point. The voltage arrow runs from plus to minus because of the technical direction of the current. The arrow is either drawn straight across the component or rounded off. Both are permissible. However, one should remain consistent and not switch between the notations.

If the start and end points are interchanged, the value of the voltage is negated.

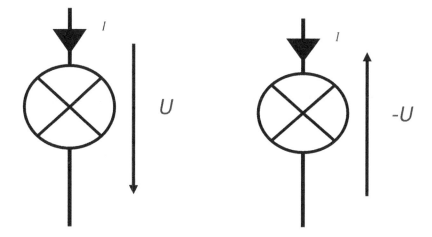

7.7 Current arrows

A current is represented by an arrow on the line. Here, too, the current is drawn from plus to minus. If the arrow points in the other direction, the value of the current is negated.

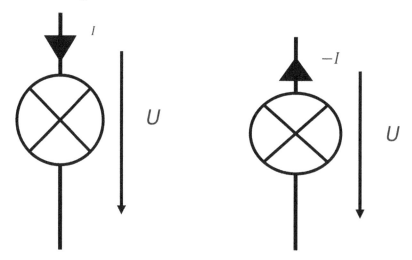

If you do not yet know in which direction the current flows, e.g., because the positive and negative poles are not yet present, one direction is assumed and calculated with this. If the calculations result in a negative current, you know that the current is actually flowing in the other direction.

Markings and treasure symbols

7.8 Producer and consumer arrow system

In electrical engineering, a distinction is made between two systems: the generator arrow system and the consumer arrow system. This is simply a matter of interpreting the voltage and current arrows. In the consumer arrow system, energy is "consumed" when current and voltage point in the same direction.

The component then absorbs electrical energy and converts it into other forms of energy.

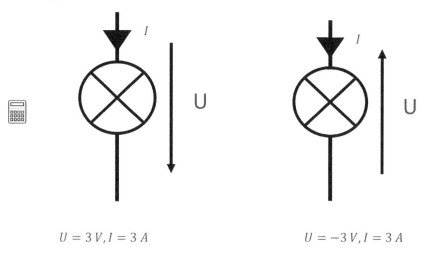

$$U = 3\,V, I = 3\,A \qquad\qquad U = -3\,V, I = 3\,A$$

If the voltage and current arrows point in opposite directions, corresponding energy is generated, for example, from a current or voltage source.

In the generator arrow system, it is the other way round. In this system, energy is "consumed" when current and voltage point in opposite directions. If they point in the same direction, corresponding energy is generated.

 In electrical engineering, the consumer arrow system is used almost without exception. The generator arrow system is only used for source observations to see whether a source is absorbing or emitting energy.

That is why the main focus in this book is on the consumer arrow system, which is used in all further examples.

7.9 Kirchhoff's laws

Next, we come to two fundamental rules that are used repeatedly in the context of circuit analysis: Kirchhoff's laws.

The physicist Gustav Robert Kirchhoff established two fundamental laws that make it easier to determine unknown voltages and currents in a circuit: the mesh theorem and the node theorem or node rule. The two laws form the basis of every circuit design or its analysis. Kirchhoff's laws are derivations from physics, and, in particular, from the law of conservation of energy. Both rules are quite logical and easy to understand.

7.10 Kirchoff law for nodes

 Wherever several lines meet, an electrical node is formed. In an electrical node, the currents are divided differently.

Put simply, Kirchhoff's first rule states that electrons or charges in an electrical node cannot simply disappear. Since a current is a moving charge, this also applies to currents. "Where electricity goes in, it must also come out again".

In physics, the node theorem more correctly states:

"In an electrical node, the sum of the currents flowing in is equal to the sum of the currents flowing out". Or: "The sum of all currents in a node is zero".

$$I_{in} = -I_{out} \qquad I_{in} + I_{out} = 0$$

In order to be able to calculate with it, the number of currents must still be determined.

 All currents running into a node are counted as positive; all currents running out are counted as negative.

The illustration shows an electrical node where a total of five lines meet.

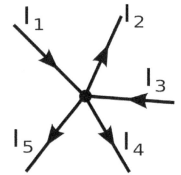

Figure 24 Electrical node

According to our definition, the flows I_1 and I_3 are positive and I_2, I_4 and I_5 are negative. The sum of the currents flowing in the nodes is therefore I_1+I_3, the sum of the currents flowing out of the node is $I_2+I_4+I_5$. The node rule states:

Markings and treasure symbols

$I_1 + I_3 - I_2 - I_4 - I_5 = 0$ or converted: $I_1 + I_3 = I_2 + I_4 + I_5$

7.11 Kirchoff law for meshes

Like the node theorem for currents, the mesh for voltages states that the sum of all voltages in a circuit must be zero.

The background here is that electrical energy must be maintained. We have already learned that power, and thus also energy, depends on the voltage. So, the voltage must also be maintained.

A mesh is a closed circuit over several voltages within a circuit.

We can choose any kind of mesh we like. The only important thing is that the starting point is the same as the end point, i.e., that a closed loop is created.

The mesh theorem states: *"The sum of all stresses within a mesh is zero"*.

Again, we need to define which tension is positive and which is negative. It does not matter whether the mesh runs clockwise or counterclockwise.

 Any tension running in the direction of the mesh is counted positively and any running against the direction of the mesh is counted negatively.

Figure 27 is an example of a mesh over two voltage sources and a consumer with the voltage U_{con}

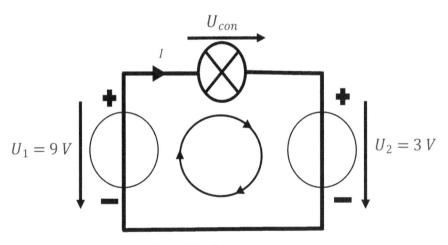

Figure 25 Mesh in a circuit

All voltages in the direction of the mesh are counted as positive, all voltages against are counted as negative.

$-U_1 + U_{con} + U_2 = 0$

$U_{con} = U_1 - U_2$

$U_{con} = 6 \text{ V}$

We have already learned about some components of the electric circuit. With the help of Kirchoff's rules, we can determine currents and voltages. In addition, we already know some components of the electric circuit such as current and voltage sources.

Of course, there are many more components that we will now go through one by one. We will start with a classic consumer, the resistor.

8 Electrical resistance

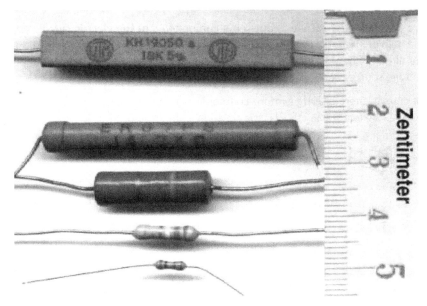

Figure 26 Different resistor designs

Electrical resistance, as the name suggests, is a resistance to electrons. It is made more difficult for the electrons to flow through the resistance. This can happen, for example, through a metal that does not give up its valence electrons "so easily". The electrons rub against each other and generate heat.

Figures 29 and 30 show the symbols used for a resistor in the US and in Europe respectively.

Figure 27 American circuit symbol Figure 28 European circuit symbol

 A resistor in the circuit is an obstacle for the current, similar to stones or pillars in the water circuit. These brace themselves against the water, thus offering it resistance.

Electrical resistance has the formula symbol R and is given in Ω (Ohm). It is named after the German physicist Georg Simon Ohm. The resistance value is a measure of how difficult it is for the current to flow through the resistance. In other words:

The electrical resistance indicates the electrical voltage U required to allow an amperage I to flow through an electrical conductor.

$$R = \frac{U}{I}$$

The unit Ohm is therefore

$$1\,\Omega = \frac{1\,V}{1\,A}$$

We can rearrange the formulas accordingly to

$$U = R \times I$$

 A mnemonic that indicates the relationship between voltage, resistance and current in a circuit is, therefore, "URI".

Excursus: The resistance of copper wire:

 Our household power cables are made of copper. The cables should conduct the current and have as little resistance as possible, but the desired resistance of 0 Ω is utopian. The resistance increases the longer the cable is and decreases the thicker the cable is, or, in other words, the larger the cross-sectional area A of the cable is.

Don't get confused, A is the formula sign for the area here, not the unit ampere.

Of course, the resistance also depends on the material. This is taken into account in the specific resistance ρ (the small Greek rho) of a material.

 The resistance of a conductor is given by $R = \rho \times \frac{l}{A}$

Copper, for example, has a resistivity of

$$\rho = 1.69 \times 10^{-2}\,\frac{\Omega mm^2}{m} \text{ to } 1.75 \times 10^{-2}\,\frac{\Omega mm^2}{m}$$

 A 10 m copper cable with a cross-section of $A = 0.1\,mm^2$ has a resistance of $R = 1.75 \times 10^{-2}\,\frac{\Omega mm^2}{m} \times \frac{10\,m}{0.1\,mm^2} = 1.75\,\Omega$

 Here, the calculation is made in square millimetres since the specific resistance is usually given in the form $\frac{\Omega mm^2}{m}$.

If this were not the case, you would first have to convert the cross-section into square metres. This would also be correct but more complicated.

Excursus: Conductance

As an alternative to the resistance R, the conductance G can be used. Conductance is a measure of how well a substance "lets through" electrons. The conductance has the unit Siemens S and is named after Werner von Siemens. The connection between conductance and resistance is relatively banal as they are reciprocals of each other.

$$G = \frac{1}{R}; R = \frac{1}{G}$$

$$1\,S = \frac{1}{1\,\Omega}; 1\,\Omega = \frac{1}{1\,S}$$

The conductance has, therefore, hardly any technical significance.

💡 In electrical engineering, one only speaks of and calculates with resistance values.

How much voltage do you have to apply to a resistor with 150 Ω for 3A to flow through it?

Solution: $U = R \times I = 150\,\Omega \times 3\,A = 450\,V$

A 9 V block battery is connected to a resistor with R = 3 Ω. How much current flows?

Solution: We provide $U = R \times I$ to look for $I = \frac{U}{R} = \frac{9\,V}{3\,\Omega} = 3\,A$

2 kA flow through a resistor while 3 MV are applied. What is the value of the resistor?

Solution: We provide $U = R \times I$ to look for $R = \frac{U}{I} = \frac{3\,MV}{2\,kA} = \frac{3 \times 10^6\,V}{2 \times 10^3\,A}$

$$= \frac{3}{2} \times 10^6\,V \times 10^{-3}\,A = 1.5\,k\Omega$$

8.1 Series connection of resistors

In electrical engineering, circuits can quickly become quite complex. Rarely is only one resistor installed, as we learned in the example. When two or more resistors are connected in a series, we speak of a series circuit.

The current must "squeeze" through both resistors one after the other. The resulting total resistance is correspondingly greater.

Individual resistors can be added together to form a single resistor. This has a resistance value equal to the sum of the individual resistance values.

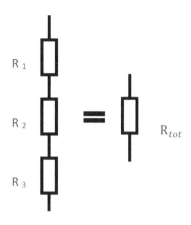

Figure 29 Series connection of resistors

$$R_{tot} = R_1 + R_2 + R_3 + \cdots$$

8.2 Voltage divider

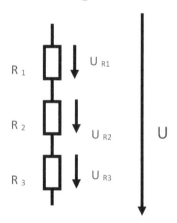

Figure 30 Voltage divider with series connection of resistors

When we have several resistors connected in series, we know the voltage drops across all the resistors together. But how much voltage is dropped across the individual resistors?

The resistors divide the voltage among themselves in proportion to their resistance values, so they form a voltage divider.

Electrical resistance

$$\frac{Voltage\ at\ resistor}{Total\ voltage} = \frac{Resistance\ value}{Total\ resistance}$$

$$U_{R1} = U_0 \times \frac{R_1}{R_1 + R_2 + R_3 + \cdots}$$

$$U_{R2} = U_0 \times \frac{R_2}{R_1 + R_2 + R_3 + \cdots}$$

...

$U_0 = 10\ V, R_1 = 6\ \Omega, R_2 = 4\ \Omega$

$$U_{R1} = 10\ V \times \frac{6\ \Omega}{6\ \Omega + 4\ \Omega} = 6\ V$$

$$U_{R2} = 10\ V \times \frac{4\ \Omega}{6\ \Omega + 4\ \Omega} = 4\ V$$

8.3 Parallel connection of resistors

If two resistors have the same starting and end point, they are connected in parallel. It becomes easier for the current to flow through both resistors than through the individual ones. The resistance value of the resulting resistor is correspondingly smaller.

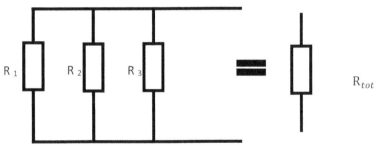

Figure 31 Parallel connection of resistors

The total resistance is calculated from the parallel connection of the individual resistors. The symbol for parallelism ∥ is used.

$R_1 \parallel R_2 \parallel R_3 \parallel \cdots = R_{tot}$

! With parallel connection, the resistance values do not add up, but the conductance values of the resistors do. $G_{tot} = G_1 + G_2 + G_3 + \cdots$

However, since we only calculate using resistance and not conductance, the result is:

$$\frac{1}{R_{tot}} = \frac{1}{R_1} + \frac{1}{R_2} + \frac{1}{R_3} + \cdots$$

8.4 Special form for two resistors

If only two resistors are connected in parallel, the formula simplifies to:

$$\frac{1}{R_{tot}} = \frac{1}{R_1} + \frac{1}{R_2}$$

If we multiply the fraction by R1 and R2 we get:

$$R_{tot} = \frac{1}{\frac{1}{R_1} + \frac{1}{R_2}} \quad => \quad R_1 \parallel R_2 = R_{tot} = \frac{R_1 R_2}{R_1 + R_2}$$

$R_1 = 10\,\Omega$, $R_2 = 30\,\Omega$

$$R_{tot} = \frac{10\,\Omega \times 30\,\Omega}{10\,\Omega + 30\,\Omega} = \frac{300\,\Omega^2}{40\,\Omega} = 7.5\,\Omega$$

8.5 Current divider

Analogous to the voltage divider rule, we are faced with the question of which current flows through resistors connected in parallel. The resistors divide the current among themselves; therefore they form a current divider.

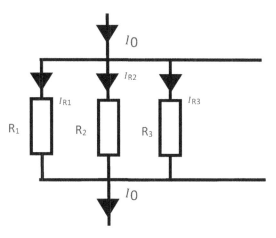

Figure 32 Current divider with resistors connected in parallel

Electrical resistance

The voltage applied to the resistors is the same. Because of "URI", this means in reverse that the currents through the resistors only depend on the resistor values R_1, R_2. The resistors divide the current inversely proportional to their resistance values or proportionally to their conductance values. We remember $G = \frac{1}{R}$.

$$\frac{Current\ through\ resistance}{Total\ current} = \frac{Conductance}{Total\ conductance}$$

💡 If one resistor is twice as large as the other, only half as much current flows through it as through the other.

$$I_{R1} = I_0 \times \frac{G_1}{G_1 + G_2 + G_3 + \cdots} \quad ; \quad I_{R2} = I_0 \times \frac{G_2}{G_1 + G_2 + G_3 + \cdots}$$

$I_0 = 10\ A,\ R_1 = 6\ \Omega,\ G_1 = \frac{1}{6\,\Omega},\ R_2 = 4\ \Omega,\ G_2 = \frac{1}{4\,\Omega}$

$$I_{R1} = 10\ A \times \frac{\frac{1}{6\,\Omega}}{\frac{1}{6\,\Omega} + \frac{1}{4\,\Omega}} = 4\ A \quad ; \quad I_{R2} = 10\ A \times \frac{\frac{1}{4\,\Omega}}{\frac{1}{6\,\Omega} + \frac{1}{4\,\Omega}} = 6\ A$$

8.6 Electrical power

We have already learnt about power and that its symbol is P. In electrical engineering, electrical power is the product of current and voltage, which is applied to a resistor, for example. $P = U \times I$

If the resistance is given instead of the voltage, the power results in:

$U = R \times I$

$P = R \times I^2$

If the resistance is given instead of the current, the power results in $P = \frac{U^2}{R}$

$$P = U \times I$$
$$P = R \times I^2$$
$$P = \frac{U^2}{R}$$

A 9 V block battery can deliver a maximum current of 0.5 A. What is the maximum power it can deliver?

Solution: $P = U \times I = 9\,V \times 0.5\,A = 4.5\,VA = 4.5\,W$

 At a resistance of 1kΩ, 10V are applied. How much power is converted in the resistor?

Solution: $P = \dfrac{U^2}{R} = \dfrac{(10\,V)^2}{1\,k\Omega} = \dfrac{100\,V^2}{1000\,\Omega} = 0.1\,W$

8.7 Application example: Resistors in a power supply unit

The following illustration shows a power supply unit with the upper housing removed. With this example we can see different components within a closed system.

 Caution: A power supply unit or other electrical components should not be opened by a layperson. Depending on their capacity, components such as capacitors can store charge for a very long time and be correspondingly dangerous.

We can recognise lots of resistors on a circuit board. These are marked on a PCB by "R", and each has its own number.

Figure 33 Internal circuit board structure of a power supply unit

Electrical resistance

9 Semiconductors: PN junctions, diodes, transistors

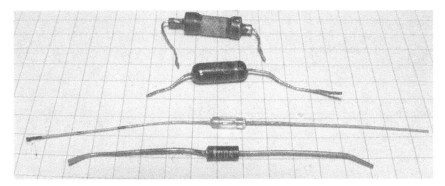

Figure 34 Different diode designs

 A diode is a component that allows current to flow in only one direction. This is why it is also referred to as a semiconductor component. In the water model, it is a kind of "cat flap" that allows water to pass through in only one direction.

Figure 35 Diode switching symbol

The diode consists of an anode, which is connected to the positive pole or higher potential, and a cathode, which is connected to the negative pole or lower potential.

 A vertical line on the component shows which wire forms the cathode.

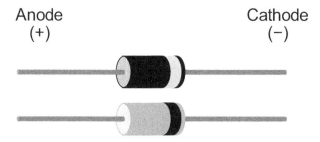

Figure 36 Anode and cathode of a diode

9.1 Structure of a diode

In the production of diodes, positive and negative atoms are doped ("implanted") into a carrier material (usually silicon). In the process, a foreign atom such as boron or phosphorus is inserted into the silicon lattice.

Silicon has four valence electrons and is neutrally charged. If a silicon atom is replaced by, for example, a boron atom with only three valence electrons, one electron is "missing". Accordingly, by substituting silicon with a phosphorus atom, which, in turn, has five valence electrons, one can create an extra electron in the material.

By inserting additional charge carriers, the material is no longer neutral but positively charged on one side and negatively charged on the other.

If there are more electrons than protons, the material is negatively charged. The material is n-doped. If there are fewer electrons than protons, the material is positively charged, p-doped. In the case of an excess of protons, one also speaks of holes or defect electrons, because where the electrons are missing, a "hole" is created.

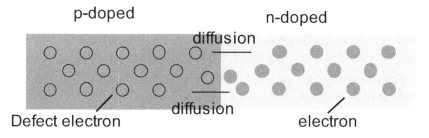

Figure 37 Structure of a PN junction

Semiconductors: PN junctions, diodes, transistors

A transition from a positive to a negative charge region is formed by the different charge carriers. This is called a p-n-junction . Exactly at the barrier layer, the charge carriers diffuse ("migrate"). The holes compensate for the excess electrons and vice versa. The technical term for this is that the charge carriers "recombine". Only neutral atoms are present in the area around the boundary layer.

The recombination of electrons and holes create a charge-free zone, the depletion zone, which continues to decrease towards the outside.

The electrons have migrated from the n-doped region to the p-doped region and the holes vice versa. This creates an electric field that prevents further charge carriers from migrating through the charge-free depletion zone

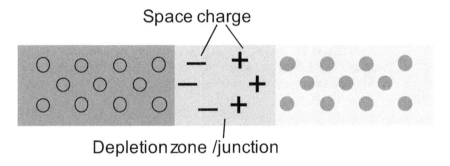

Figure 38 Space charge zone of a PN junction

If a positive voltage is applied to the n-doped side and a negative voltage to the p-doped side, the electrons are sucked off even further, and the depletion zone increases. Thus, no current can flow at all. The other way around, things look better:

If a positive voltage is applied to the p-doped side and a negative voltage to the n-doped side, additional electrons and holes are pushed into the charge zones and "flood" the depletion region. As a result, the current can now flow unhindered from the positive pole of the voltage to the negative pole.

For this reason, diodes are often used to protect against reverse polarity or **overvoltage**. If the voltage becomes too high, the diode starts to conduct. The diode breaks down and diverts the current. Another application is to limit the current in one direction, for example, when charging a battery and preventing it from discharging.

The fact that the diode only conducts in one direction means that the diode can be operated either in the flow direction or in the reverse direction.

The direction in which the diode conducts can be seen from the switch symbol; the arrow indicates the direction of the current in which the diode conducts.

 If the diode is conductive, the current must pass through the PN junction. The PN junction itself acts as a voltage source that reduces the potential.

The voltage that must be applied at the PN junction to "flood" the depletion zone depends on the type of diode.

At the PN junction of a standard silicon diode, approximately 0.7 V are needed. With a Schottky diode, for example, it is only 0.2 V. There are other types of diodes, for example, the Zener diode, also known as a breakdown diode. This diode blocks in both directions but breaks down above a certain applied voltage (breakdown voltage).

 A diode has almost no resistance in the direction of flow and therefore does not limit the current flow. You need another resistor, otherwise there is the danger of a practical short circuit.

In the power supply unit below (figure 35.2), we can also find many diodes. They are used, for example, for voltage conversion from the changing input voltage to a constant voltage. The diodes are marked with D and a number. Optically they do not differ significantly from a resistor.

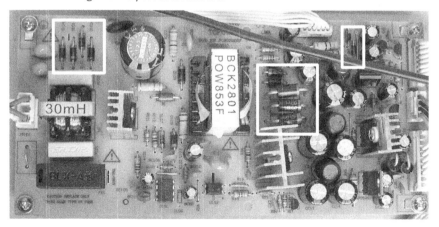

Figure 35.2 Internal circuit board structure of a power supply unit

Semiconductors: PN junctions, diodes, transistors

9.2 Excursus: LED

Light-emitting diodes, or, simply LEDs have become an indispensable part of our everyday lives.

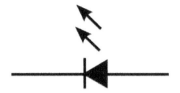

Figure 39 LED switching symbol

The mode of operation and properties of a light-emitting diode are the same as those of a "normal" PN semiconductor diode. An LED has a flow direction and a reverse direction, hardly any resistance in the flow direction and consists of a carrier material that has been n- or p-doped. The big difference is the carrier material used. While "normal" diodes are made of silicon, LEDs usually use a gallium compound as a semiconductor material. In addition, LEDs usually have a higher forward voltage of approximately 1.6 V - 3.6 V instead of 0.7 V, as is the case with a non-luminous diode.

 A diode also has almost no resistance in the direction of flow and therefore does not limit the current flow. A wide resistance is needed, otherwise there is a risk of a practical short circuit. This is called a **series resistor**.

Depending on the colour of the LED, the forward voltage also differs.

A white LED has a forward voltage of approximately 2.8-3.2 V, a red or green LED usually only 2.0-2.3 V. The voltages depend on the type of construction and the semiconductor used. The LED manufacturer specifies a range for the permissible forward voltage.

Accordingly, the series resistor must be adapted to the colour.

9.3 The transistor

Figure 40 Different transistor designs

Transistors are a very common component. Today, **more than 10 billion** transistors are built into the Intel processor of a standard PC. With the help of transistors, **arithmetic operations** can be realized, which are the basic building block of every **digital system**.

 A transistor can be regarded as a **controllable electrical switch**. This allows the current to pass through or blocks it completely.

The development and integration of transistors in **industrially manufactured chips** (e.g., in a processor for a PC) has been the most significant development in recent decades.

It initiated **digitalization** and the **automation** associated with it. In a book on electrical engineering, transistors should therefore not be missing under any circumstances.

Nowadays, transistors are no longer connected by hand, but **substrate** or **bulk is** etched into a carrier material. As with diodes, the substrate forms the **silicon**. The smallest transistors are in the range of a **few nanometres** and are only a few atoms in size!

 Transistors are divided into two major categories:
Bipolar transistors and **field effect transistors**.

Semiconductors: PN junctions, diodes, transistors

Their mode of action is very similar, but the physical effects behind them are fundamentally different. The **bipolar transistor** is used not only as a switch but also as a **current amplifier**, whereas the field-effect transistor is used almost without exception as a switch.

9.4 The bipolar transistor

A bipolar transistor is obtained by connecting two diodes with opposite polarity. This creates three differently charged zones.

Depending on how the diodes are connected, two n-doped zones and a p-doped zone in the middle (NPN transistor) or two p-doped zones and an n-doped zone in the middle (PNP transistor) are created.

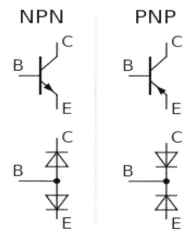

Figure 41 Structure and circuit symbol of an NPN and PNP transistor

The **collector C** is located where the current flows in. The operating voltage is applied to the collector. The **base B is located in the** middle. The "output" of a transistor is called the **emitter E**.

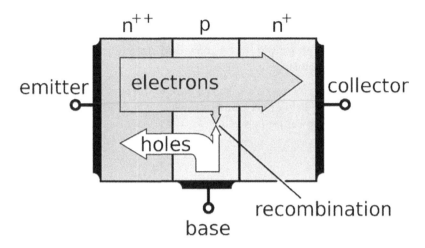

Figure 42 Structure of an NPN transistor

Figure 44 shows an **NPN transistor**. The collector and base are **n-doped** n+ and n++ respectively. This means that they are **particularly strongly n-doped**.

💡 In the off-state, the transistor is not conductive, after all, it consists of two blocking diodes.

By applying a **positive voltage** to the base, the electrons are pulled from the emitter to the base.

💡 The depletion area is flooded with electrons, and a current can flow from the collector to the emitter. **The transistor is conductive.**

The PN junction or NP junction, which is actually non-conductive, becomes conductive so that the electrons **from the collector** can flow through the two PN/NP junctions. The electrons then flow **through the emitter** towards the lower potential. The name of the **bipolar transistor** comes from the fact that both positive and negative charge carriers are involved in the "current transport".

In the case of the transistor, the current is not limited in the **on-state** from the base to the emitter since the PN junction has no resistance in the direction of flow.

An additional **resistor, a base series resistor,** is needed **to limit the current**, otherwise you risk an enormously high current. As with the diode, an additional 0.7V drops between the base and the emitter in the conductive state.

💡 An NPN transistor becomes conductive when a voltage of at least 0.7 V is applied between base and emitter.

In reality, a transistor does not become immediately conductive. From a base-emitter voltage of **about 500 mV,** it slowly starts to suck in electrons. At a base-emitter voltage of 0.7 V, it is completely switched through. The **collector and emitter** are virtually at the **same potential** in the conductive state. The voltage from the collector to the emitter in the connected state is about 0.2 V and is called the **saturation voltage.**

The problem of limiting the current from the base to the emitter (base current) is circumvented by the **field effect transistor**, which is fundamentally different in design and yet has similar properties.

9.5 The field effect transistor

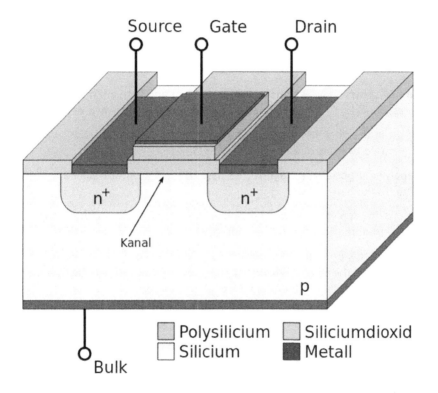

Figure 43Structure of a field effect transistor

The field effect transistor, FET for short, works fundamentally differently from a bipolar transistor. As the name suggests, a field, namely the electric field, is used here to switch the transistor on and off.

The terms of the connections in the field-effect transistor are also different. Instead of the collector, the upper entry point is called the source S, the base corresponds to gate G, and the emitter is replaced by drain D. The source and drain are each n-doped, i.e., they have an excess of electrons.

💡 The gate is separated by a thin insulating layer of silicon dioxide. It is, therefore, not electrically connected to the rest of the transistor.

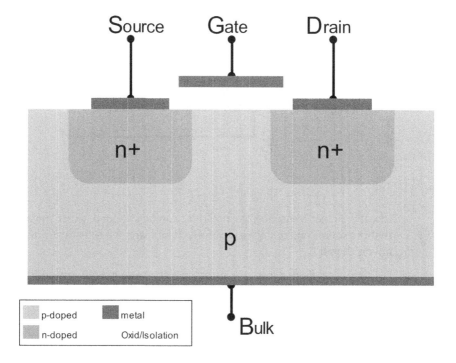

Figure 44 Structure of an N-FET

If a positive voltage, i.e., an excess of protons, is applied, an electric field is created from the gate to the p-doped substrate. The electrons are "sucked in", and a conductive channel forms between the drain and the source.

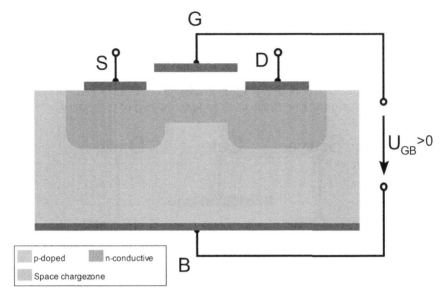

Figure 45 Structure of an N-FET

 Like the bipolar transistor, the polarities can be swapped, which means that the source and drain are positively doped, and the substrate is negatively doped.

FETs are also divided into transistors, which are switched on or off by default.

 A positive channel is then formed in the on-state. Accordingly, one speaks of an n-channel or p-channel FET.

Another design is the MOSFET (metal-oxide-semiconductor field effect transistor).

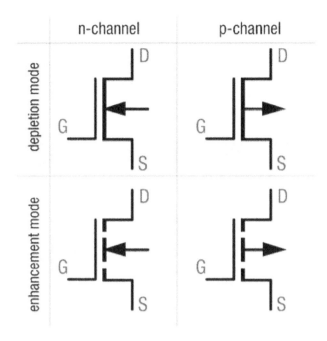

Figure 46 Switching symbols n-channel and p-channel

The power supply unit below (Figure 35.3) also contains transistors, primarily MOSFETs as switching transistors. They switch on and off thousands of times per second and can thus transform a high input voltage down to the required output voltage. Since loss occurs during switching, which is expressed in heat, the transistors are screwed to a heat sink.

Figure 35.3 Internal circuit board structure of a power supply unit

Semiconductors: PN junctions, diodes, transistors

10 The capacitor

Figure 47 Different capacitor designs

The capacitor is a very frequently used component and can be found several times in every circuit. It is a passive component, so it does not need a power supply. Furthermore, like a battery, it has the ability to store electrical charges, but only for shorter periods of time. In the water model, a capacitor would be an additional water basin that can absorb and release a lot of water for a short time. In this way, it can compensate for inconstant water inflows.

Various circuit symbols for capacitors

Figure 48 Circuit symbols for different capacitor types

There are many different capacitors: ceramic capacitors, electrolytic capacitors, variable capacitors or trimming capacitors. The structure of these components is basically the same and quite simple to understand.

 They consist of two electrically conductive surfaces, the electrodes, that face each other but are separated by insulation, the dielectric.

Figure 51 illustrates the plate capacitor – the simplest in terms of construction. Two flat plates are mounted opposite each other:

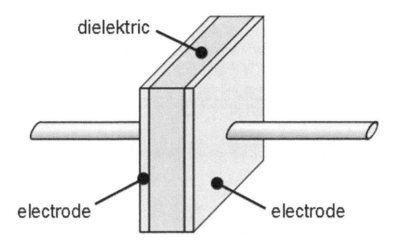

Figure 49 Structure of a plate capacitor

The connections or plates are called electrodes. If a voltage is applied to the capacitor, charge carriers reach the plates. This causes an electric field to form.

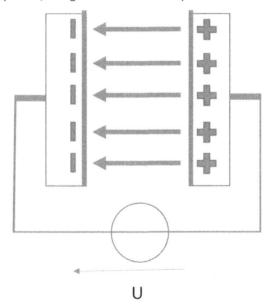

Figure 50 Structure of the homogeneous E-field in the plate capacitor

If the voltage is increased, more charge carriers are attracted to the plates, and a correspondingly higher charge Q is stored on the plates.

The capacitor

The charge stored on the plates is therefore dependent on the voltage U applied. This is a proportional relationship between voltage and charge, which can be expressed by a proportionality constant. This constant is different for each capacitor and is determined by the design, size, material, and many other factors.

 The constant that indicates the relationship between voltage and charge is called capacitance C. Its unit is the farad F.

$Q = U \times C$

To determine the capacity, we convert the formula according to the capacity C.

$C = \dfrac{Q}{U}$

$1\,F = \dfrac{1\,C}{1\,V}$

 Capacity is a measure of the storage of electrical charge, not energy. Colloquially, however, this distinction is often not made.

For a plate capacitor, the capacitance is easy to calculate by the design. For this we make two assumptions:

1. The larger the plate areas, the more charge can be stored at the same voltage. The capacity is proportional to the plate area d $C \sim A$.

2. The further apart the plates are, the less charge can be stored at the same voltage. The capacity is inversely proportional to the plate distance d $C \sim \dfrac{1}{d}$

The formula is supplemented by a natural constant of the electric field ε_0 as well as a constant of the dielectric ε_r.

$C = \varepsilon_0 \times \varepsilon_r \times \dfrac{A}{d}$

 ε_0 is called the electric field constant and has the numerical value of $8{,}85 \times 10^{-12} \,\dfrac{As}{Vm}$.

ε_r is called relative permittivity and depends on the material. In a vacuum or in air $\varepsilon_r = 1$. Paper has an ε_r of one to four, water of about 80 and good insulators up to over 10,000. In reality, the capacitances of capacitors are relatively small. The magnitude of the capacity is in the micro-, nano- or picofarad range.

10.1 Charging a capacitor

We have already learned that a capacitor can store charges and thus electrical energy. If you connect a capacitor to a voltage source, charge carriers are stored on the plates. In the following, we will look at how long a capacitor takes to charge or discharge and how much energy can be stored.

All the following derivations for the capacitor are made to facilitate basic understanding. The electro-technical effects always occur when charging an electrical storage device, regardless of whether it is a simple plate capacitor in the circuit, the battery from the latest smartphone or the 75 kWh battery of an e-car. The modes of operation, charging and discharging curves are analogous to the simple plate capacitor.

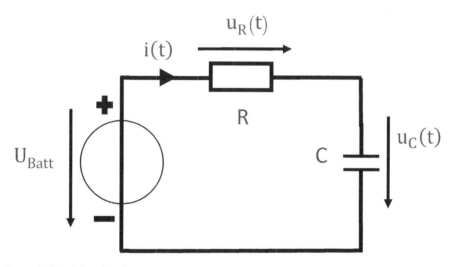

Figure 51Circuit for charging a capacitor

To understand the charging processes, let's assume the simplest circuit with a capacitor, a voltage source, here a simple battery, and a resistor. We will see later why we absolutely need the resistor.

 This combined circuit of capacitor and resistor is also called an RC element.

We are looking for the functions $u_C(t)$, and $i(t)$ which depend on the time t and describe the capacitor voltage and the current in the circuit during the charging process.

Why is the electric current named $i(t)$ and not $i_C(t)$?

The capacitor

Theoretically $i_C(t)$ would also be correct, but there is only one current in the entire circuit. The same current flows through the resistor and the capacitor, so differentiation by indexing is not necessary.

Derivations

To do this, let's take another close look at the circuit and apply Kirchhoff's second rule, the mesh theorem. We run through a mesh over all three voltages $u_R(t), u_c(t)$ and U_{Batt}.

$$U_{Batt} - u_R(t) - u_c(t) = 0$$

$$U_{Batt} = u_R(t) + u_c(t)$$

Using this equation, we now go through three points in time or time full stops during charging.

At the time $T_0 = 0\,s$, the power supply is connected, and the capacitor charges.

1. $T_1 = T_0 = 0\,s$

The capacitor is still completely "empty" and can hold many charge carriers. The voltage applied to the capacitor is zero because there is no charge yet.

$$U_c = \frac{Q}{C} = 0$$

$$U_R = U_{Batt}$$

 At this moment, the capacitor is not a resistor for the current. The current flows as if the capacitor were not there.

The current is then given by $I = \frac{U_R}{R} = \frac{U_{Batt}}{R}$

 We also see immediately why the resistor is necessary. It limits the charging current at the beginning of the charging process. Without the resistor, there would be a practical short circuit. In the worst case, the capacitor explodes, or the voltage source fails.

2. $T_0 < t < \infty$

Over time, the capacitor becomes charged. More and more charge carriers are pressed onto the plates. However, not as many charge carriers as desired can be charged onto the plates because the charge carriers already on the plates repel those that follow. So fewer and fewer new charge carriers flow in; accordingly, the current continues to decrease, and the capacitor voltage steadily increases.

3. $t \to \infty$

If we wait long enough, the plates of the capacitor are full of charge carriers. The voltage source cannot press any more charge carriers onto the plates. The current here is accordingly zero. Since no more current flows, no more voltage drops across the resistor.

$U_R = I \times R$

The entire voltage is now applied to the capacitor.

$U_{Batt} = U_c$

$I = 0$

 The capacitor blocks the flow of charge carriers, so it is an infinitely large resistor for the circuit.

You would have to increase the voltage of the battery to push additional charges onto the capacitor.

 To get the function that gives us the capacitor voltage and current in the circuit, we once again apply our mesh formula:

$U_{Batt} = u_R(t) + u_c(t)$

Here, we replace

$u_R(t) = i(t) \times R$

and

$u_{c(t)} = \frac{q(t)}{C}$

and get

$U_{Batt} = i(t) \times R + \frac{q(t)}{C}$

We also know that the current represents the charge per time interval. As a differential, we write $I = \frac{dQ}{dt}$

The dt stands for an infinitesimal, temporal change. We get what is called a differential equation.

$U_{Batt} = \frac{dq(t)}{dt} \times R + \frac{q(t)}{C}$

We can leave the solving of the differential equation to mathematicians, for us, only the derivation and the solution are important.

If we solve the differential equation according to q(t) and replace the charge by current and voltage, we get a function for the voltage and the current at the capacitor. The solution of the differential equation is:

$$u_C(t) = U_{Batt} \times \left(1 - e^{\left(-\frac{t}{R \times C}\right)}\right)$$

$$i_C(t) = \frac{U_{Batt}}{R} \times e^{\left(-\frac{t}{R \times C}\right)}$$

We see that an **e – function** describes the charging process of the capacitor and that charging takes longer if the capacitor has a higher capacity, as it can then store more charge carriers.

In addition, the charging process takes longer when the resistance is greater, as it is an obstacle for the charge carriers and reduces the current flow.

> The product of capacitance C and series resistance value R is, therefore, also called the time constant τ (tau). τ = RC

Finally, we plot the voltage and the charging current over time. The time is given in multiples of the time constant since the charging curve depends on it.

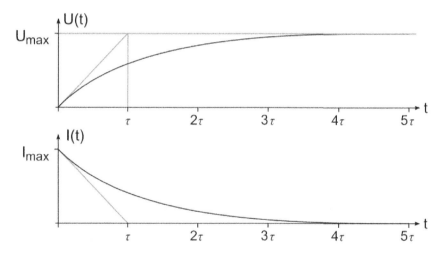

Figure 52 Charging curve of a capacitor

$$U_{max} = U_{Batt}; I_{max} = \frac{U_{Batt}}{R}$$

> Tau does not describe how long the capacitor needs until it is fully charged!

The capacitor

Example: If τ = 5s, the capacitor is not fully charged after 5 seconds, but only to approximately 63.7%. ($0.637 = e^{-1}$). After 10 s (=2 τ) it is about 86% ($0.86 = e^{-2}$) after 15 s (3 τ) to almost 95% and after 25 s (5 τ) it is almost 99.3% charged.

Theoretically, a capacitor is never 100% charged. In practice, it is usually quite sufficient to say that a capacitor is fully charged after more than 5 τ.

10.2 Discharging the capacitor

The discharging of a capacitor is analogous to the charging process. For this, we again assume the simplest circuit with a capacitor and a resistor. At the beginning of the discharging process, the capacitor is charged to the level of the battery voltage. $U_C = U_{Batt}$. Then the voltage supply is disconnected and replaced by a short circuit (a piece of wire). This can be realised, for example, by a switch that alternates between battery voltage and a short circuit.

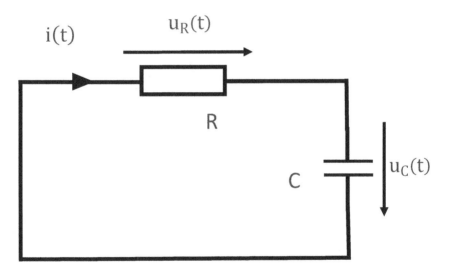
Figure 53 for discharging a capacitor

We are again looking for the functions $u_C(t)$ as well as $i_C(t)$ which depend on the time t, and the capacitor voltage as well as the current in the circuit during the discharge process. The mesh equation gives us:

$u_R(t) - u_c(t) = 0$

$u_R(t) = u_c(t)$

In this one we replace $u_R(t) = i(t) \times R$ and $u_{c(t)} = \frac{q(t)}{C}$

$$i(t) \times R = +\frac{q(t)}{C}$$

We substitute $I = \frac{dQ}{dt}$ and get a differential equation.

$$\frac{dq(t)}{dt} \times R = +\frac{q(t)}{C}$$

Again, we will skip deeper mathematical approaches and content ourselves with the solution of the differential equation.

$$u_C(t) = U_{Batt} \times e^{\left(-\frac{t}{\tau}\right)}$$

$$i_C(t) = -\frac{U_{Batt}}{R} \times e^{\left(-\frac{t}{\tau}\right)}$$

With $\tau = RC$ and the initial value of the capacitor voltage of U_{Batt} we plot the voltage and the charging current over time:

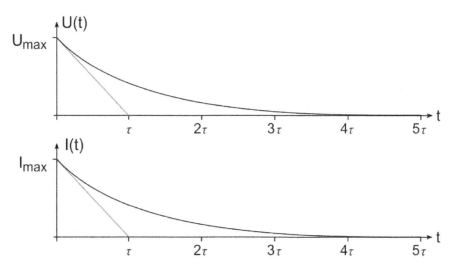

Figure 54 Discharge curve of a capacitor

$$U_{max} = U_{Batt}; I_{max} = -\frac{U_{Batt}}{R}$$

! Attention: The current flows out of the capacitor and is therefore negative. The graph only shows the amount of the current!

Again: Theoretically, a capacitor is never 100% discharged. In practice, it is usually quite sufficient to say that the capacitor is discharged after 5τ.

Excursus: Charging batteries and rechargeable batteries

In principle, batteries are nothing more than very large capacitors. Although batteries store energy chemically and not electrically, the charging and discharging processes are similar.

Therefore, when charging batteries and accumulators, one would also have to use a resistor to limit the current. However, the resistor would only convert precious electrical power into unused heat.

A considerable amount of energy would be lost each time the battery is charged. That is why batteries are not charged from a fixed voltage source but from an intelligent charger. This charger, for example the 5V power supply for our smartphone, limits the current so that the battery is not damaged. This means that the current can be limited without a charging resistor.

10.3 How much energy can a capacitor store?

We know how many charges a capacitor with capacity C can hold, namely
$$Q = U \times C$$
But how much energy is stored in the capacitor when fully charged?

The energy is stored in the electric field of the capacitor. The amount of energy can be calculated as follows
$$E = \frac{1}{2} \times Q \times U$$

With
$$Q = U \times C$$

we get
$$E_{el} = \frac{1}{2} \times C \times U^2$$

How long does a capacitor, with a capacity of 100 nF and a charging resistance of 1 kOhm, need until it is fully charged? What is then the current that flows into the capacitor when 9 V is applied?

Solution: After about five time constants, the capacitor is charged. $5\tau = 5 \times RC = 5 \times 1000\Omega \times 100\ nF = 500\ \mu s$
The current has then become zero.

How long does the same capacitor take to discharge?
Solution: Just as long as when charging, i.e., 500 µs

Why do you need a series resistor for the capacitor when charging?
Solution: To limit the maximum charging current.

How much energy can a capacitor with 200 µF store at a voltage of 4 kV?

Solution:

$$W_{el} = \frac{1}{2} \times C \times U^2$$

$$= \frac{1}{2} \times 200 \, \mu F \times (4 \, kV)^2$$

$$= \frac{1}{2} \times 200 \times 10^{-6} \, F \times 16 \times 10^6 \, V^2$$

$$= 1600 F \, V^2$$

$$= 1.6 \, kJ$$

10.4 Application area of capacitors

We have seen that capacitors are very good at storing charges, but only for a short time. This makes capacitors perfect for supporting voltages and currents.

If a voltage source cannot provide the required amount of current, the capacitor serves as a buffer storage. It releases a lot of energy in a short time when needed. Afterwards, when there is little load, it recharges.

 Such backup capacitors are almost always installed in parallel with supply voltages to support them.

For example, in power supply units, on circuit boards or household appliances. They protect against overvoltage peaks or if the supply voltage briefly collapses, for example due to a large load jump at the consumer.

Furthermore, capacitors are used to filter high frequencies, for example in applications such as data processing or audio amplification.

In the power supply unit below (figure 35.4), we find a lot of capacitors that support both the input voltage and the output voltage. Furthermore, numerous capacitors are installed to filter interference voltages.

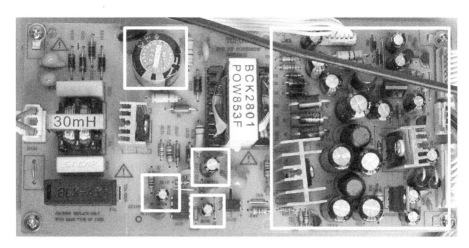

Figure 35.4 Internal circuit board structure of a power supply unit

The capacitor

11 The coil

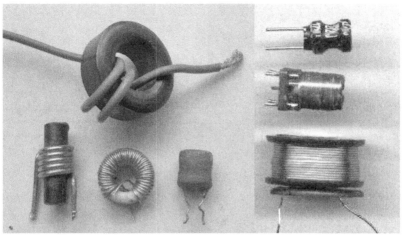

Figure 55 Different coil designs

In addition to the capacitor, the coil is also a very frequently used component and can be found several times in every circuit. However, coils are significantly more expensive, both in terms of material and production. That is why circuit designers try to minimise this component and replace it with capacitors or resistors if necessary.

Just like the capacitor, the coil is a passive component. It has the ability to store electrical energy.

However, it does not use an electric field for this, but a magnetic field. Again, the energy can only be stored for a short time. The coil can be drawn in a circuit diagram in two different ways:

Figure 56 Circuit symbols of a coil

 A coil is nothing more than a wire that is wound several times around a body.

Usually, the coil is wound on a material with good magnetic conductivity, such as iron or ferrite. However, air coils are also possible, i.e., coils without a winding body.

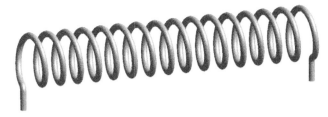

Figure 57Construction of an air coil

The simplest construction is a cylindrical coil. A wire is wound onto a cylinder with a round cross-section. In Figure 59, the core has been removed, so it is an air core coil.

When a current flows through the coil, a magnetic field is created. The magnetic field that is formed depends on the current that flows through the coil.

 Analogous to the capacitance of the plate capacitor, the characteristic quantity of a coil is its self-inductance, or simply inductance L.

This indicates how "well" the coil can build up a magnetic field. In colloquial language, we therefore sometimes speak of an inductance instead of a coil. The unit of inductance is Henry H, named after the American mathematician Joseph Henry.

$$1\,H = 1\,\frac{kg \times m^2}{A^2 \times s^2}$$

The inductance is different for each coil and is determined by the design, size and many other factors.

With a cylindrical coil, the inductance is relatively easy to calculate.

$$L = \mu_0 \times N^2 \times \frac{A}{l}$$

Here μ_0 is the constant of the magnetic field.

The coil

The formula is an approximation equation that applies to long cylindrical coils. In practice, the value of the coil is written on the component. With self-winding coils, this formula can therefore be used.

In reality, the inductances of coils are relatively small. The order of magnitude of the capacitance is in the milli, micro, nanohenry range.

In our water model, the coil corresponds to a paddle or a flywheel with a large mass. In contrast to the consumer, it does not serve to draw energy from the circuit but runs passively with the water flow. Due to its high mass, however, it is very inert. At the beginning, the water must first push the paddle wheel. So, it is braked until the paddle wheel starts to move. If the pump is switched off afterwards, the wheel continues to run for a certain time because of the inertia and continues to drive the water. We see that the water flow cannot start or stop abruptly. The paddle wheel would continue to stop or push the water.

11.1 Magnetic coupling

Another property of coils is the transfer of energy from one coil to another. Let us imagine a second water circuit for this purpose. In each circuit there is a flywheel, and both flywheels are connected to each other by a shaft. If one flywheel is driven, the other is automatically driven and sets the second water circuit in motion. Energy is transferred from one circuit to the other via the shaft without the water flows being connected.

In the electric circuit, the magnetic field represents the "wave". For a magnetic coupling, two coils are wound on a common core.

If a current flows through one of the coils, a magnetic field is built up. This induces an induction voltage and a current flow in the second coil. This is referred to as a transformer. In a transformer, energy is transferred from one coil to another without contact.

The power supply unit below again shows the familiar power supply unit. The circled coil is wound on a ferrite core and serves as an input filter. In the middle, we see a transformer framed in a square, which consists of two coils wound into each other.

Figure 35.5 Internal circuit board structure of a power supply unit

11.2 Switch-on process of a coil

To understand the charging of coils, we must first consider what properties a coil has when a voltage is applied. To do this, we apply our water model. We already know that the coil can be thought of as a paddle wheel with a very large mass.

When the pump starts working, the wheel blocks the water. It begins to rotate slowly but continues to slow down the water. In the process, the paddle wheel absorbs energy.

Analogously, the coil in the electric circuit throttles the current flow.

 Since the coil throttles the current flow, a coil is also called a current restriction .

Gradually, the paddle wheel gets going and turns as fast as the water flows. It is almost no longer an obstacle to the water cycle.

We have already learned that a coil generates a magnetic field and can thus store energy. If you connect a voltage source to the coil, a current starts to flow.

In the following, we will look at how long a coil needs to build up the magnetic field completely or, in other words, to be fully charged.

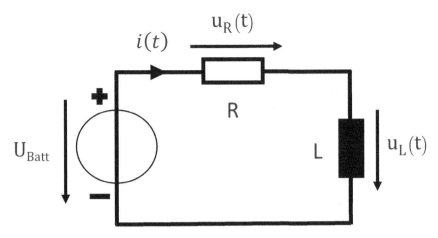

Figure 58Circuit for 'charging' a coil

All the following derivations for the coil are made to facilitate a basic understanding of the charging process. For this purpose, the simplest circuit with a coil, a resistor and a voltage source, here a simple battery, is assumed.

💡 This combined circuit of coil and resistor is also called an RL element.

Again, we look for the functions $u_L(t)$, and $i(t)$ which depend on the time t and thereby describe the coil voltage and the coil current in the circuit during the charging process.

Derivations

To do this, let's take another close look at the circuit and apply Kirchhoff's second rule, the mesh theorem. We run a mesh through all three voltages $u_R(t), u_L(t)$ and U_{Batt}.

$$U_{Batt} - u_L(t) - u_R(t) = 0$$
$$U_{Batt} = u_L(t) + u_R(t)$$

Using this equation, we can plausibly explain three points in time or time intervals.

At the time $T_0 = 0s$ the voltage is applied, and the coil charges.

1. $T_1 = T_0 = 0s$

At the time 0s, many effects take place that were explained in the chapter on electromagnetism. If it is difficult to understand the following processes, it is advisable to look them up again.

The voltage is applied, and the current begins to flow.

A magnetic field builds up in the coil.

The structure of the magnetic field corresponds to the change in magnetic flux. This means that a voltage is U_{ind} induced.

$$U_{ind} = -\frac{d(B \times A)}{dt}$$

According to Lenz's rule, this voltage counteracts its cause, the magnetic field increases (current increase).

As a result, no current flows at time 0s, and no voltage drops across the resistor. $U_{Batt} = U_L = -U_{ind}$

2. $T_0 < t < \infty$

Over time, the magnetic field builds up. More and more energy is stored. The current in the circuit increases accordingly, the voltage $u_L(t)$ decreases and the voltage $u_R(t)$ increases.

3. $t \to \infty$

If we wait long enough, the maximum current value is reached. The magnetic field has been completely built up. The full voltage drops across the resistor. U_{Batt} drops. $U_{Batt} = U_R$

No voltage is induced on the coil anymore, and the coil voltage has become zero. $U_L = 0$.

The current in the circuit is limited accordingly by

$I = \frac{U_{Batt}}{R}$.

To get the function that gives us the coil voltage and current in the circuit, we add our mesh formula:

$U_{Batt} = u_L(t) + u_R(t)$

With this one we replace $u_R(t) = i(t) \times R$

Analogous to the capacitor voltage $u_{C(t)} = \frac{q(t)}{C}$ the coil voltage is dependent on the excitation current, which changes over time.

$$u_{L(t)} = L \times \frac{di(t)}{dt}$$

The derivation of this equation has been deliberately omitted because for us, the analogy to the capacitor voltage is sufficient. We obtain a differential equation again:

The coil

$$U_{Batt} = L \times \frac{di(t)}{dt} + R \times i(t)$$

As usual, mathematics provides us with the solution to the differential equation:

$$i_L(t) = \frac{U_{Batt}}{R} \times \left(1 - e^{\left(-\frac{R \times t}{L}\right)}\right)$$

$$U_L(t) = U_{Batt} \times e^{\left(-\frac{R \times t}{L}\right)}$$

We see that again a e − Function describes the charging process of the coil.

The quotient of resistance and inductance forms our time constant τ (tau). $\tau = \frac{L}{R}$

All the properties of the capacitor, such as the estimate that the coil is fully charged after 5τ, are also valid here.

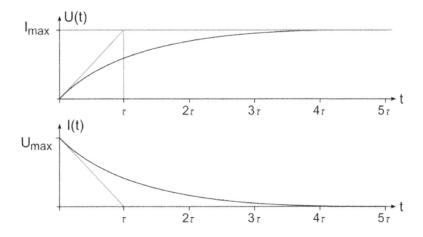

Figure 59 Charging curve of a coil

$$U_{max} = U_{Batt}; I_{max} = \frac{U_{Batt}}{R}$$

11.3 Switching off a coil

Switching off a coil is analogous to switching it on. For this, we again assume the simplest circuit with a coil, a switch and a resistor. At the beginning, the current of the circuit is maximum, and the coil voltage is $U_L = 0$.

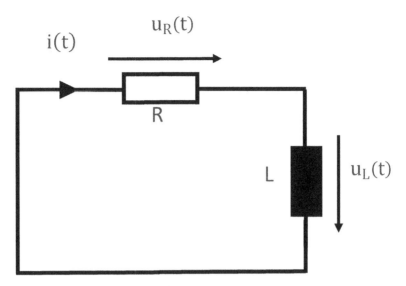

Figure 60 Circuit for 'discharging' a coil

We are again looking for the functions $u_L(t)$ and $i_L(t)$ which depend on the time t and thereby describe the coil voltage and the current in the circuit during the demagnetisation process. A complete mesh circulation results in

$$u_R(t) - u_L(t) = 0$$
$$u_R(t) = u_L(t)$$

We replace $u_R(t) = i(t) \times R$ and $u_{L(t)} = L \times \frac{di(t)}{dt}$

And get the differential equation again.

$$i(t) \times R = L \times \frac{di(t)}{dt}$$

The solution is:

$$i_L(t) = \frac{U_{Batt}}{R} \times e^{\left(-\frac{t}{\tau}\right)}$$

$$u_L(t) = -U_{Batt} \times e^{\left(-\frac{t}{\tau}\right)}$$

With $\tau = \frac{L}{R}$ and the initial value of the coil current $I_{max} = -\frac{U_{Batt}}{R}$

Let's plot the voltage and charge current over time:

$$U_{max} = U_{Batt}; I_{max} = -\frac{U_{Batt}}{R}$$

The coil

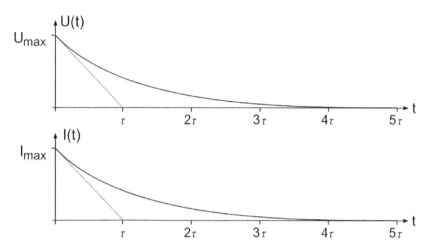

Figure 61 Discharge curve of a coil

 Attention: The voltage at the coil is negative from the moment it is switched off, as it counteracts the cause (current decrease).

$$U_{ind} = -\frac{d(B \times A)}{dt}$$

The graph shows the amount of voltage.

11.4 How much energy can a coil store?

The amount of electrical energy that a coil can store in its magnetic field depends on the current flowing through the coil and the coil's self-inductance. The energy is stored in the electric field of the capacitor. The amount of energy can be derived as

$$E = \frac{1}{2} \times H \times B \text{ or } W_{mag} = \frac{1}{2} \times L \times I^2$$

How long does a coil with an inductance capacity of 300 µH and a charging resistance of 2 kΩ need until it is fully charged? What is the current when a block battery with a voltage of 9 V is applied?

Solution: After about five time constants the coil is charged.

$$5\tau = 5 \times \frac{L}{R} = 5 \times \frac{300 \text{ µH}}{2000 \text{ Ω}} = 750 \text{ ns}$$

$$I = \frac{U}{R} = \frac{9 \text{ V}}{2 \text{ kΩ}} = 4.5 \text{ mA}$$

How much energy can a coil store with an inductance of 400 µH with a charging resistance of 1 kΩ and an applied voltage of 9 V?

$$W_{mag} = \frac{1}{2} \times L \times I^2$$

$$= \frac{1}{2} \times 400\ \mu H \times 0.9\ A^2$$

$$= 162\ \mu J$$

11.5 Comparison capacitor and coil

We have already seen that capacitors and coils have many things in common. The following table shows the most important characteristics.

Component	Capacitor C	Coil L
Formula symbol and unit	Farad F	Henry H
Energy storage	Electric field	Magnetic field
Energy in the component	$W_{el} = \frac{1}{2} \times C \times U^2$	$W_{mag} = \frac{1}{2} \times L \times I^2$
Field of application	Voltage and current stabilisation filters	Current smoothing Filters Transformers
Time constant τ during unloading/loading	$\tau = R \times C$	$\tau = \frac{L}{R}$
Resistance in uncharged state	Zero	Infinite
Resistance in charged state	Infinite	Zero

12 Practical example - LED switch-on delay

We have worked out numerous theoretical basics. We already know many components and their modes of action. We are ready to examine a small circuit and understand its mode of operation.

The aim of the circuit is to switch on an LED, not immediately, but with a **time delay**. To do this, we need a supply voltage, for example a 9 V block battery, a transistor, an LED, a series resistor for the LED and a base resistor for the transistor, which also forms the charging resistor for the capacitor.

Purpose	Component	Designation/value
Voltage source	Battery	9 V block battery
Switch	NPN transistor	BC548C
Light source	LED	5 mm LED white
Series resistor	Resistance	380 Ω
Base resistance	Resistance	100 kΩ
Time delay	Capacitor	220 µF

The LED and the series resistor R2 are connected in series and are connected to the supply voltage and the collector C of the transistor T. U_{Batt} and the collector C of the transistor T.

Resistor R1 and capacitor C are also connected in series and are connected between the supply voltage and earth. The emitter E of the transistor T is directly connected to ground.

12.1 The circuit

The circuit is constructed as follows:

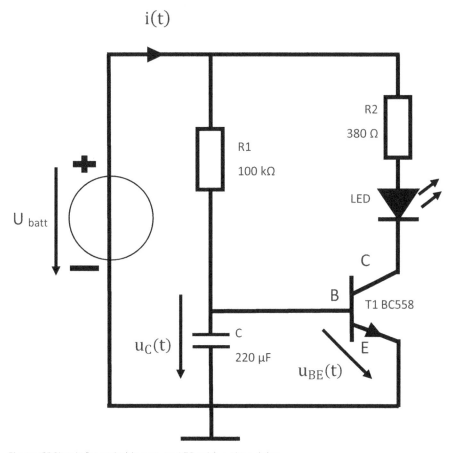

Figure 62 Circuit for switching on an LED with a time delay

We draw a very simple mesh across the capacitor voltage and the base-emitter voltage of the transistor.

$$u_{BE}(t) - u_C(t) = 0 \text{ bzw. } u_{BE}(t) = u_C(t)$$

The capacitor voltage $u_C(t)$ is equal to the base-emitter voltage $u_{BC}(t)$ of the transistor.

What happens when the supply voltage of 9 V is connected?

The power supply is applied. The capacitor is completely empty. The capacitor voltage is therefore after $U_C = \frac{Q}{C} = 0$.

We remember that the transistor does not switch through until the base-emitter voltage becomes greater than 0.7 V. Therefore, transistor T does not conduct in the beginning, and the LED does not light up.

Over time, the capacitor is charged via the resistor R1, more and more charge carriers reach the capacitor, and the capacitor voltage increases.

In this state, the capacitor is charged with the known charging curve (e-function), but the transistor continues to block, and the LED cannot light up.

The voltage at the capacitor continues to rise until 0.7 V is reached. Now the transistor can switch through so that a current can flow from the 9 V supply voltage via the LED and the resistor to earth. The LED thus lights up after a time delay.

12.2 Calculating the time delay

The time delay depends directly on the charging curve of the capacitor.

The transistor switches through as soon as the capacitor voltage is raised above $u_{BE}(t) = u_C(t) = 0.7\ V$ is. We have already derived the formula for the capacitor voltage.

$$u_C(t) = U_{Batt} \times \left(1 - e^{\left(-\frac{t}{R \times C}\right)}\right)$$

So, the following must apply to the capacitor voltage

$$0.7\ V = U_{Batt} \times \left(1 - e^{\left(-\frac{t}{R \times C}\right)}\right)$$

$$\frac{0.7\ V}{9\ V} = 1 - e^{\left(-\frac{t}{R \times C}\right)}$$

$$\frac{0.7\ V}{9\ V} - 1 = -e^{\left(-\frac{t}{R \times C}\right)}$$

$$-0.922 = -e^{\left(-\frac{t}{R \times C}\right)}$$

$$\ln(0.922) = -\frac{t}{R \times C}$$

$$t = -\ln(0.922) \times R \times C$$

The time delay results from the choice of resistor and capacitor.

For our values of $R = 100\ k\Omega$, $U_{Batt} = 9\ V$, und $C = 220\ \mu F$ we obtain a switch-on delay of:

$$t = -\ln(0.922) \times 100\ k\Omega \times 220\ \mu F \approx 1.79\ s$$

By varying the resistor R1 and the capacitor C, the time can be shortened or lengthened.

The circuit is then built using a breadboard. Figure 65 shows the finished, discretely constructed circuit. The switch is only used to switch the supply voltage on and off.

Can you identify the remaining components? Which component is the capacitor, which is the transistor, and which is the LED or the resistor?

Figure 63Discrete circuit layout on a breadboard

Practical example - LED switch-on delay

13 Introduction to alternating current theory

All the assumptions, calculations and examples we had made so far assumed that voltages, currents or power were constant. So far, this has also been largely accurate.

For example, the 9 V battery permanently supplies a constant 9 V. The potential of the positive pole is 9 V "higher" than that of the negative pole.

If we only consider quantities that do not change over time, we generally speak of **direct current** abbreviated as **DC**.

However, it is not always the case that the voltage or other variables remain constant. The easiest way to understand why this is the case is to take a look at how electricity can be generated.

Disclaimer:
Alternating current theory is a very complex subject. In order to be able to fully comprehend all the interrelationships, several years of study are necessary. Therefore, the following sections are by no means intended to be complete. Some areas, such as the pointer theory or areas from higher mathematics, such as the complex numbers, have been deliberately simplified or reduced to the most necessary basics. The focus is clearly on the understanding and significance of the theory in practice. Therefore, real-life examples are introduced.

13.1 Power generation

We have already done a lot of calculations with electricity and voltage. But how is electricity provided in the first place?

There are many ways to generate a voltage or a current flow.

Lightning is probably the oldest phenomenon in which electricity plays a role. In lightning, different poles are formed. The lightning we perceive during a thunderstorm is so-called cloud-to-earth lightning. In this case, a strongly negative pole forms within a cloud in relation to the earth's surface.

 Nearly all lightning (80-90%) is not cloud-to-earth lightning but cloud-to-cloud lightning. This occurs when charges separate at different altitudes. The lightning forms the equalising charge from one cloud to another. We perceive the cloud-to-earth lightning, if at all, as the sky lighting up.

At some point, the voltage is high enough to ionise the air molecules. This means that electrons and protons within a molecule are separated by the strong electric field.

However, this also creates a conductive channel because we have learned that electricity is nothing more than moving charge carriers.

Through this channel, the lightning discharges with a current strength of up to more than 100 kA.

The effect that lightning is caused by natural electricity was already confirmed by Benjamin Franklin in 1752. He flew a kite in a thunderstorm and thereby provoked a lightning strike.

Theoretically, it is possible to harness the energy from lightning. However, since lightning occurs unevenly, it does not make economic sense to use this natural phenomenon to generate electricity.

The first real use of electricity was discovered by an English inventor whose name we have already encountered several times. We are talking about Alessandro Volta, after whom the unit of voltage was also named.

Volta used the chemical property of metals for this purpose: When two metals come into contact with each other, the less noble metal always dissolves. In this context, dissolving means that it oxidises and decomposes, an example of which is the rusting of copper.

This creates a voltage that until then could neither be measured nor used.

Around 1800, however, Volta invented a way to demonstrate this effect. He took metal plates made of zinc and copper and stacked them on top of each other. Between them were leather rags soaked in a salt solution. Zinc, which is chemically the less noble metal, dissolved and gave off electrons. The zinc atoms dissolved in the salt solution, but the electrons remained. As a result, the zinc plate formed the negative pole of the first battery cell.

The voltage could be increased by connecting several of these Volta cells in series. In this case, connected in series meant nothing more than that the cells were stacked on top of each other. The zinc plate at the lower end was the negative pole; the copper plate at the upper end was the positive pole.

This gave rise to the first usable battery, the **Volta Column**.

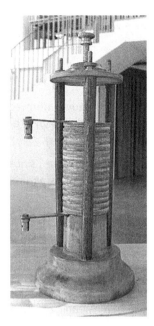

Figure 59 Multi-stage Volta column

We see that electricity has been used for over 200 years. But what does electricity generation look like nowadays?

A large part of electricity generation now comes from renewable energies. These include solar plants and wind power.

The principle of a solar plant system is relatively simple to understand

We have already learned about the effect of a PN transition. As a reminder, let's use the graphic of a PN transition again.

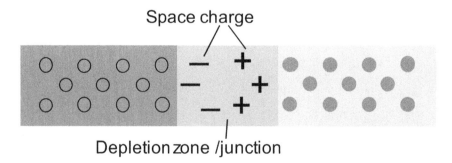

Figure 60 p-n-junction

In principle, a solar plant system is nothing more than a large-area PN junction, i.e., silicon doped with boron and phosphorus.

This gives us a silicon wafer that has many free electrons. We cannot use them yet, because they are firmly bound to the nucleus.

However, when high-energy solar rays hit the doped silicon, the free electrons are separated from their nucleus. The electrons are "knocked out", a voltage is created, and the electrons can flow away via a consumer.

In practice, a single solar cell generates a DC voltage of 0.5-0.6 V.

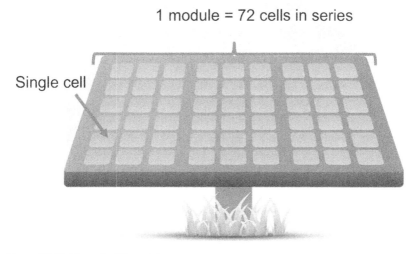

Figure 61 Division of a PV module

Usually, 48, 60 or 72 of these cells are connected in series so that a solar panel has an output voltage in the 30 V range.

Solar cells generate a **DC voltage** in the process.

The situation is different, however, when we generate electricity from **electrical generators.** A generator converts a rotary motion into an electrical voltage.

Introduction to alternating current theory

13.2 Power generation by means of generators

A miniature generator that almost everyone knows is the classic bicycle dynamo. This generates electricity from the rotation of the bicycle tyre, which is used to power the light. But what is the physical background? How can a movement be converted into electricity?

All we need is a magnet with a permanent magnetic field. Either a permanent magnet or a current-carrying coil can be used for this. Because, as we know, a current-carrying conductor also generates a magnetic field.

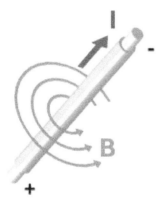

Figure 62 Magnetic field of a current-carrying conductor

To simplify the derivation, we again use a permanent magnet. We mount it on an axis so that we can rotate it.

Next to this magnet, we set up a coil. When we start to rotate the magnet on its axis, the magnetic field that passes through the coil changes.

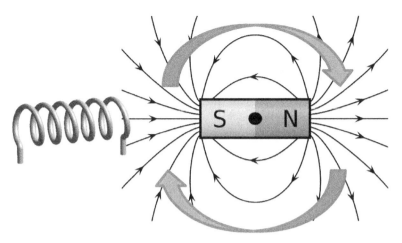

Figure 63 A permanent magnet rotates next to an air coil

We further recall that a changing magnetic field always entails an induction voltage.

$$U_{ind} = -\frac{d(B \times A)}{dt}$$

This means that when a magnet rotates next to the coil, an electrical voltage is induced in the coil. A voltage can be measured between the ends of the coil.

However, this voltage is not uniform but depends on the position of the magnet. This is because the magnetic field of the permanent magnet is also not homogeneous.

The change in the magnetic field is highest when the magnet is in the perpendicular position to the coil. Analogously, the change is smallest (namely exactly zero) when the magnet is horizontal to the winding axis of the coil. In addition, it must be noted that the sine of the magnetic field also changes!

The following table summarises this:

Introduction to alternating current theory

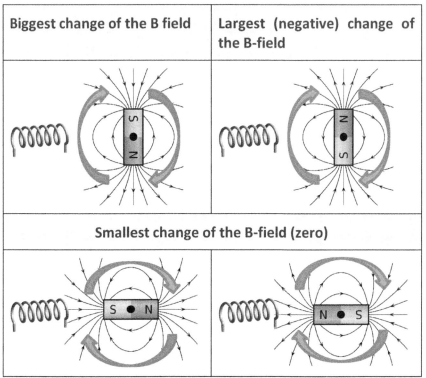

Figure 64 A permanent magnet rotates next to an air coil

The voltage at the ends of the coil is not linear but represents the circular motion of the rotating magnet. The image of this circular movement, therefore, results in a sine curve. The sine changes after every half turn.

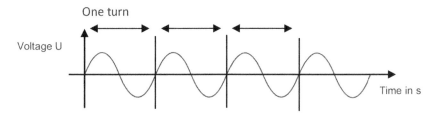

Figure 65 AC-voltage over time

This voltage changes the sine. This is why this type of voltage is also called alternating voltage. It is often referred to as alternate current, AC for short.

Before we look at the exact areas of application and conclusions, we need to introduce a few uniform terms.

The maximum voltage value measured at both the positive and negative peaks is called the **peak value**, **maximum voltage** or **amplitude** and is marked with a circumflex (^).

 A complete revolution is called a period, and the time it takes to complete it is called the period time. It is abbreviated with a *T and* indicates the time it takes for the magnet to return to its initial position. The unit of the period is the second.

For a "normalised sine or cosine oscillation", the period is $T = 2\pi$ (see chapter 2.8. Sine, Cosine, Tangent).

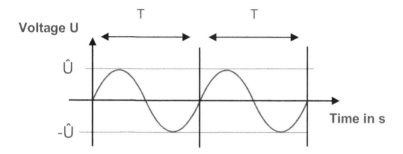

Figure 65 AC-voltage over time

The reciprocal of the period is the frequency f. This indicates how many periods occur within one second.

 The unit of frequency is named after the German physicist Heinrich Hertz. $[f] = Hz$ named.

Since the frequency indicates how often an oscillation occurs per second, the frequency is just the reciprocal of the period duration and vice versa.

$f = \frac{1}{T}; T = \frac{1}{f}.$

The unit Hz is therefore equal to $\frac{1}{s}$. As an alternative to the frequency f, the **angular frequency** ω is often used. The angular frequency refers to a "normalised" cosine oscillation of 360° or 2π written as a radian.

$\omega = 2\pi f$

2π is a numerical value of ~6.28, therefore the unit of the angular frequency is also $\frac{1}{s}$.

One hertz corresponds to $\frac{1}{s}$, but the unit Hertz Hz *is reserved* exclusively for the frequency f. The angular frequency is therefore given in $\frac{1}{s}$,never in Hertz Hz.

If the magnet rotates once through 360° within one second, this corresponds to a frequency of $f = 1\ Hz$ the angular frequency is $\omega = 2\pi f = 6.28\ \frac{1}{s}$.

 A magnet makes three revolutions within one second. What is the frequency, the angular frequency and the period of the resulting sine wave?

Solution:

$$f = \frac{3\ turns}{1\ s} = 3\frac{1}{s} = 3\ Hz$$

$$\omega = 2\ x\ \pi\ x\ f = 2\ x\ \pi\ x\ 3\ Hz = 18.85\ \frac{1}{s}\ (not\ Hz!)$$

$$T = \frac{1}{f} = \frac{1}{3\frac{1}{s}} = 0.33\ s$$

 The German electricity grid is operated at a frequency of 50 Hz. What is the period duration of the grid?

Solution: $T = \frac{1}{f} = \frac{1}{50\frac{1}{s}} = 0.02\ s = 20\ ms$

After having learned the most important quantities within a sinusoidal oscillation, we can represent the oscillation as a concrete formula. We get the momentary induced voltage between the ends of the coil depending on the position of the magnet or the time.

$$u(t) = \hat{U} \times \sin(2\pi \times f \times t)$$

The peak value of**Fehler! Textmarke nicht definiert.** the sinusoidal voltage corresponds to the maximum induction voltage. This results from, amongst other things, the strength of the magnet and the number of coil windings.

At this point, we can save ourselves the time-consuming calculation of the peak value. It is important to understand that we can generate an alternating voltage with the help of a rotating movement of a magnet and a coil. This is output as a sinusoidal voltage.

However, there are still some special features that an AC voltage brings with it. For this, we consider what the **average voltage of** the AC voltage is. The average voltage corresponds to the *time average value*, which is often abbreviated as $\overline{U}$

Since the voltage is not constant, the average value changes over time. Therefore, we look at the course over an entire period.

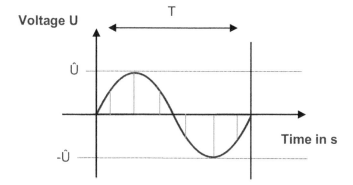

Figure 66 AC-voltage over time

We see that the sine curve has an average voltage value of zero. This is because every voltage part above the X-axis is neutralised by a voltage divider below the X-axis.

$\overline{U} = 0$

Does this mean that we cannot use the AC voltage to operate a lamp, for example? After all, the average voltage is zero.

Although the average voltage is zero, the decisive variable for operating electrical devices is the energy or power that is transmitted.

The power is made up of the product of current and voltage. This means that even with an average voltage of 0 V, energy can still be transmitted.

To illustrate this, let's take our simple generator and connect a resistor to it.

Introduction to alternating current theory

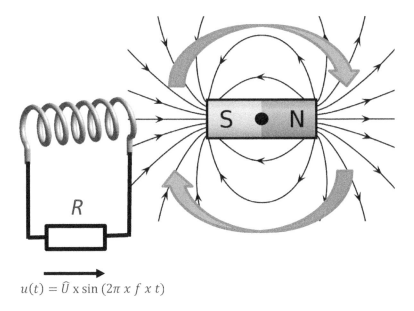

$$u(t) = \widehat{U} \times \sin(2\pi \times f \times t)$$

Figure 64A permanent magnet rotates next to an air coil to which a resistive load is connected

The power results from the resistance to

$$P = \frac{u(t)^2}{R} = \frac{\left(\widehat{U} \times \sin(2\pi \times f \times t)\right)^2}{R}$$

We see that the power shows a sine-square curve. This has an average value that is greater than zero.

Instead of complicated calculations with mean value and power, we use a trick. We use a voltage value that converts the same average power at a resistive load, i.e., at a simple resistor. This average value is also called the **root mean square value** or **the effective value of the sinusoidal voltage.**

 The **effective value is** also called the **RMS value** (Root-Mean-Square value). The RMS value is not limited to sinusoidal functions. It can be used to compare any type of function.

The function is recorded over the time interval of a complete period, squared and then the root of the time average is formed.

$$U_{Eff} = \sqrt{\frac{1}{T}\int_0^T u(t)\,dt}$$

Introduction to alternating current theory

We can insert our sine wave into the formula and then simplify it numerically. After several transformations, a very simple solution emerges:

For a sinusoidal oscillation, the effective value is given by

$$U_{Eff} = \frac{1}{\sqrt{2}} \times Û \approx 0.707 \times Û$$

This means that a sinusoidal AC voltage with the function

$$u(t) = 100\ V \times \sin(2\pi \times f \times t)$$

converts the same power over time as a DC voltage with a voltage value of

$$U = \frac{1}{\sqrt{2}} \times Û \approx 0.707 \times Û = 70.7\ V$$

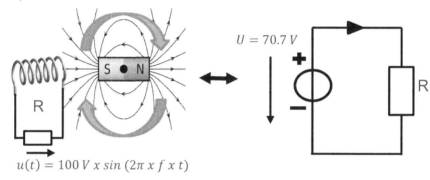

Figure 68 Converting the AC voltage source to a DC voltage source

Without going deeper into the derivation and the associated integral calculus, the following table shows which RMS values different voltage curves reveal.

Introduction to alternating current theory

RMS values of periodic voltage signals

Voltage curve	Mean value	RMS value U_{Eff}
Sinus: $u(t) = \hat{U} \times \sin(2\pi \times f \times t)$ [sine wave graph with amplitude $\hat{U}$, period T]	$\bar{U} = 0$	$U_{Eff} = \dfrac{1}{\sqrt{2}} \times \hat{U}$
[cosine wave graph with amplitude $\hat{U}$, period T] Cosinus: $u(t) = \hat{U} \times \cos(2\pi \times f \times t)$	$\bar{U} = 0$	$U_{Eff} = \dfrac{1}{\sqrt{2}} \times \hat{U}$
[DC voltage graph at level $\hat{U}$] DC voltage: $u(t) = \hat{U}$	$\bar{U} = \hat{U}$	$U_{Eff} = \hat{U}$

Introduction to alternating current theory

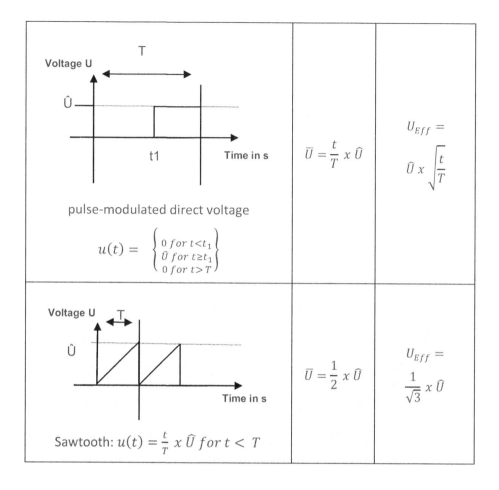

To become more familiar with RMS values, let's try some exercises:

 A generator produces a uniform sinusoidal voltage with a peak value of $\hat{U} = 400\ V$. What is the RMS value of the voltage?

Solution:
$$U_{Eff} = \frac{1}{\sqrt{2}} \times \hat{U} = \frac{1}{\sqrt{2}} \times 400\ V = 283\ V$$

 Our power grid supplies a sinusoidal voltage curve with an effective value of $U_{Eff} = 230\ V$. How much voltage can be measured in the peak? How much power is converted in a 1000 Ω resistor when we connect it to the mains?

Solution:

Introduction to alternating current theory

$$U_{Eff} = \frac{1}{\sqrt{2}} \times \hat{U}$$

$$\hat{U} = \sqrt{2} \times U_{Eff} = \sqrt{2} \times 230\,V = 325\,V$$

$$\frac{U^2}{R} = \frac{U_{Eff}^2}{R} = \frac{(230\,V)^2}{100\,\Omega} = 52.9\,W$$

 What is the peak value of a sawtooth voltage that has the same effective value of $U_{Eff} = 230\,V$? How much power is converted in the same 1000 Ω consumer?

$$U_{Eff} = \frac{1}{\sqrt{3}} \times \hat{U}$$

$$\hat{U} = \sqrt{3} \times U_{Eff} = \sqrt{3} \times 230\,V = 398\,V$$

The same power (52.9 W) is converted in the resistor since the effective value is the same for both voltage curves.

We have seen so far that a generator at its core consists of nothing more than a coil that induces a voltage through a changing magnetic field. This induced voltage follows a sinusoidal course. The voltage has no average value, but we can transmit power. To be able to calculate the power, we use the effective RMS-value .

In real generators, several pairs of coils are arranged in a circle. In order for the magnetic flux to be distributed more evenly, the coils are wound on ferrite cores.

The rigid construction is called the stator.

A rotor is inserted into this stator with magnets on the outside. When the rotor turns, it generates an induction voltage in the coils connected to it. There are different designs of generators. For example, sometimes the coils are built into the rotor and rotate with it, while the permanent magnets are fixed in the stator.

There are other ways of constructing generators, for example by using additional pairs of coils instead of permanent magnets, which in turn energise themselves. These coils are called excitation coils. The advantage of this is that the strength of the current within the exciter coils can be adjusted to determine how much power the generator should produce. It also saves the cost of permanent magnets, which are the most expensive components of a generator. The disadvantage of a so-called separately excited generator is that the excitation coils themselves produce losses and control electronics may be required.

Before we get to the effects of alternating current on various components, let's take a look at a practical example, namely how our power grid is constructed.

13.3 Structure of the power grid

It has already been mentioned that the electricity grid in Europe has an effective value of 230 V and a frequency of 50 Hz. If we measure the voltage between the poles of a socket, we get a sinusoidal voltage with an effective value of 230 V and a frequency of 50 Hz. For the average household, this is perfectly adequate. Depending on the cable cross-section, it can deliver power in the range of up to 3.7 kW.

For the supply of entire cities or municipalities, however, it is extremely awkward to work with a sine wave voltage with an RMS value of 230 V. For large outputs, very thick cables are needed.

To solve this problem, two tricks are used. The first is simple. The required thickness of a cable depends primarily on the current that flows through it, not on the voltage. In order to be able to transport the same power with a smaller cable cross-section, the voltage can therefore be increased.

Example based on a cable cross-section of 1.5 mm²	
Maximum load capacity of the cable: 15 A	
RMS value of the voltage	**Maximum transmissible power**
$U_{Eff} = 230\ V$	$P_{Max} = 230\ V \times 15\ A = 3.4\ kW$
$U_{Eff} = 400\ V$	$P_{Max} = 400\ V \times 15\ A = 6\ kW$
$U_{Eff} = 1\ kV$	$P_{Max} = 1\ kV \times 15\ A = 15\ kW$

Based on this simple consideration, we can see that it makes sense to raise the voltage, especially for transmission over long distances. This reduces the current required and the cable cross-sections for the transmission of the same power.

That is why the electricity grid is divided into different voltage levels.

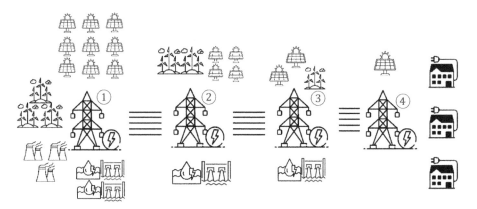

Figure 69 Structure of the European electricity grid at different voltage levels

Voltage level	Use	RMS value
① Extra-high voltage level	Directly from large generators e.g., coal-fired power plants, wind farms, hydropower plants	220 kV / 380 kV
② High voltage	Medium-sized plants such as large solar parks, medium pumped storage etc.	60 kV - 110 kV
③ Medium voltage	Smaller power generators, individual wind farms, smaller solar farms, gas-fired power plants	6 kV - 30 kV
④ Low voltage	Small generators like a PV system on the roof	230 V / 400 V

By dividing it into different voltage levels, it is possible to transport electrical energy over several hundred kilometres.

This is another reason why we use an AC grid – high voltages are needed to transmit long distances. AC voltages are much easier to transform to high voltages than DC voltages. Put simply, all we need for a transformer that transforms a low AC voltage into a higher AC voltage are two coils with different windings.

Furthermore, the generators that are used for most of the electricity generation today produce an alternating voltage. This can be transformed via transformers and then transmitted over long distances. This saves losses during voltage increase compared to a direct current grid.

Another measure that brings even more advantages than just minimising losses is one that does not only transmit one sinusoidal oscillation but three. These are not identical, however, but **shifted** in **phase**. Let's take a closer look at these terms and their effects:

The **phase, phase shift**, phase difference or phase position of a sine wave is abbreviated with the Greek letter Phi φ. It indicates the temporal displacement of the wave in relation to another wave.

Here 360° or 2π corresponds to a whole period. Therefore 90°, for example, corresponds to a quarter period. A shift to the left corresponds to a negative phase shift, a shift to the right to a positive phase shift.

$u(t) = \hat{U} \times \sin(2\pi \times f \times t)$ (black)

$u(t) = \hat{U} \times \sin(2\pi \times f \times t - 90°)$ (grey/blue)

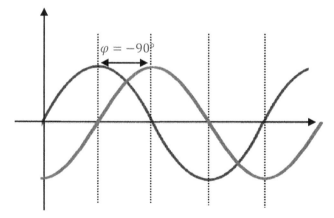

Figure 65 Phase shift of -90°

Three oscillations, called phases, are also used in the power grid. These phases are called **L1, L2**, and **L3 and** are evenly shifted. The first phase L1 has no phase shift. $\varphi = 0°$, L2 has a phase shift of $\varphi = 120°$ and the phase L3 has a phase shift of $\varphi = 240°$. This means that all 3 phases are evenly shifted against each other. These have a common **zero point N** (neutral conductor) as a **reference point.**

Introduction to alternating current theory

Each phase individually has a voltage **U1, U2** and **U3**. In a house connection, which is assigned to the low-voltage network, the phases have an effective value of the already known 230 V compared to the neutral conductor.

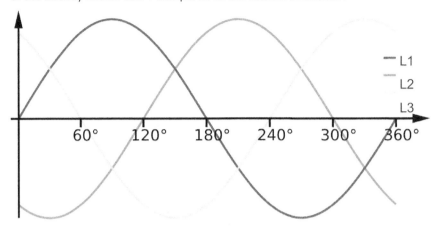

Figure 66Three-phase alternating current

In addition to the three phases and the neutral conductor, a **PE protective conductor is** often used. This is directly connected to the earth potential and serves as protection against contact, as we already know from direct current technology.

Three-phase alternating current is also called **three-phase current** or sometimes **power current.** The colour of the sheathing is standardised for the phases and the neutral and earth conductors. This means that an electrician can tell from the colour of the cable what function the cable has and whether voltage is potentially present. The standardisation is as follows:

Designation	Function	Colour
L1	Phase 1 $\varphi = 0°$	Brown
L2	Phase 2 $\varphi = 120°$	Black
L3	Phase 3 $\varphi = 240°$	Grey
N	Neutral conductor	Blue
PE	Protection/earth conductor	Green-yellow

DIN VDE 0293-308 (VDE 0293 Part 308):2003-01 and HD 308 S2

This standardisation applies both in Germany and throughout the European Union.

A five-core three-phase cable, such as is found in home electronics, therefore contains exactly five cores.

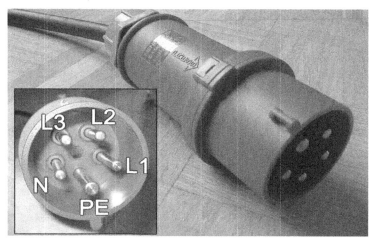

Figure 67Power plug with five connections

The three phases are provided by the electricity supplier and then connected to the house distribution box. From there, the phases are separated, and each phase is individually routed to the required rooms.

That is why the ordinary Schuko socket only has three connections. Two pins that carry one phase (L) and one neutral (N), as well as two protective contacts (PE) at the top and bottom of the plug.

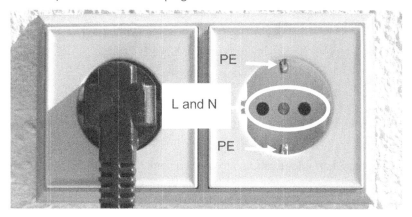

Figure 68Pin assignment of a Schuko socket outlet

Introduction to alternating current theory

Appliances that require a lot of power, for example an induction cooker or an oven, often use all three phases at the same time.

We now know that three-phase alternating current is used in the European power grid. But why use just three and not four, five or ten phases?

This question is resolved when we look at some relationships that arise from the three-phase alternating current.

This time we are not looking at the voltage between one phase and the neutral, but at the voltage between two phases, for example, between L1 and L2. Since both phases are AC voltage, the difference is **time-dependent**:

$u_{L_1}(t) = \hat{U} \times \sin(2\pi \times f \times t)$

$u_{L_2}(t) = \hat{U} \times \sin(2\pi \times f \times t - \mathbf{120\,°})$

$u_{L_1 L_2}(t) = \hat{U} \times \sin(2\pi \times f \times t - \mathbf{120\,°}) - \hat{U} \times \sin(2\pi \times f \times t)$

The differential voltage is time-dependent. At certain times the differential voltage is exactly zero, at other times, it is at its maximum. We can also see this from the voltage curves. The arrows represent the differential voltage.

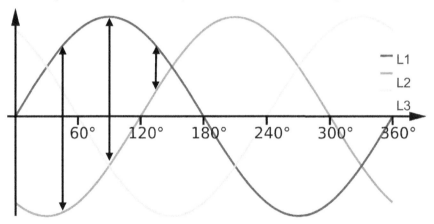

Figure 69 Illustration of the differential voltage between two phases

But how can we then calculate with the time-dependent differential voltage $u_{L_1 L_2}(t)$?

We have already learned the solution to this problem. Since the differential voltage repeats periodically, we use the RMS value of the differential voltage.

This can be calculated numerically via the definition of the effective value. The result is simple. The result is an effective value of:

$$U_{L_1 L_2\_Eff} = \sqrt{3} \times U_{L1\_N\_Eff} = \sqrt{3} \times U_{L2\_N\_Eff}$$

The RMS value of the conductor-to-conductor voltage is greater than the RMS value of the individual conductors with respect to the neutral conductor by a factor of root 3.

Since both RMS values of the phases L1 and L2 are identical, they are usually not distinguished in the designation.

$$U_{L1\_N\_Eff} = U_{L2\_N\_Eff} = U_{Eff} = U$$

Furthermore, in three-phase systems one speaks almost exclusively of effective values. Therefore, the suffix "Eff" is often omitted. This results in the simplified designations:

$$U_{L_1 L_2} = \sqrt{3} \times U$$

The factor $\sqrt{3}$ is also called the **concatenation factor**. This relationship always applies when three identical phases shifted by 120° are used.

As a beginner, there are now several sizes that go together. As these are quite similar, it is easy to confuse them. Therefore, we will summarise the most important sizes:

U_{Eff} / U: RMS value of a phase with respect to the neutral conductor.
Example low voltage network: 230 V

$\hat{U}$: Maximum value of the sinusoidal voltage of *a phase*. This is greater by the factor $\sqrt{2}$ greater than the effective value.
Example low voltage network: 325 V

$U_{L_1 L_2\_Eff}$ / $U_{L_1 L_2}$: RMS value of the voltage between two phases. This is (in a three-phase system) greater by a factor of $\sqrt{3}$ than the r.m.s. value of a single phase with respect to the neutral conductor.
Example low voltage network: 400 V (Exact 398 V)

---

 the voltages of a three-phase network are named after the effective value of the interlinked conductor-conductor voltage. The low-voltage grid is therefore also called a *400 V three-phase grid*.

Having clearly distinguished the terms, let us look at why exactly three phases were chosen. The reason lies in the temporal course of the system's power. The power of a simple sinusoidal voltage varies over a period. In the zero crossing of the voltage, for example, the power after $p(t) = u(t) \times i(t)$ is also zero.

Introduction to alternating current theory

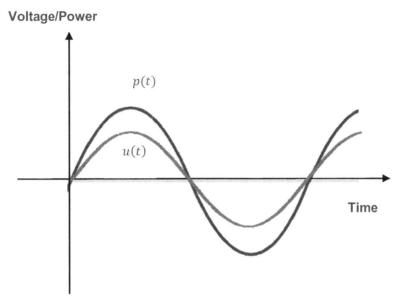

Figure 70 Voltage and power curve with a two-phase system

However, this is not the case with a three-phase system.

 In a three-phase system, the power is divided between all phases. As a result, the line that we can draw from the three-phase grid is also constant at all times.

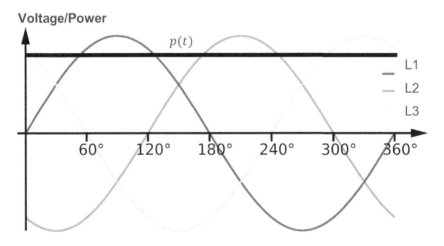

Figure 71 Voltage and power curve for a three-phase system

This physical relationship is possible *from three phases.* Theoretically, a four-phase system with four sine waves, each shifted by 90°, would also be possible. However, the additional copper required for the cables would be disproportionate to the additional benefit.

A constant provision of power over a full voltage period is necessary, especially for electrical machines. With the help of a three-phase network, a uniform, magnetic field can be generated to accelerate the machine.

An analogy is provided with the internal combustion engine. This uses several pistons to generate a largely constant torque during a "petrol combustion period".

So far, we have concentrated strongly on macroelectronics and the construction of power grids.

We have seen that a generator generates a sinusoidal alternating voltage and that the German power grid is divided into three phases. But what property and, above all, what effects does this alternating voltage have on our previously known components such as the capacitor or a coil?

14 Components in the AC circuit

So far, we have learned about the components and their behaviour at a constant voltage. This is why we also speak of the **direct current behaviour of** the components. We also already know the circuit symbol for a direct current source. Analogously, the circuit symbol of an alternating current source contains a "tilde" or a sine wave.

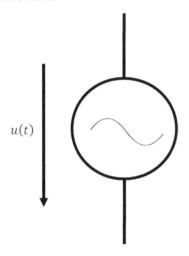

Figure 72Switching symbol of an alternating current source

The symbol is often used, but it is not a standard. Alternatively, the symbol of a direct current source can be used, and the function written on it.

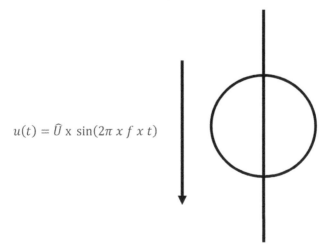

Figure 73Alternative circuit symbol of an alternating current source

However, we will see that the behaviour of components such as a coil or a capacitor in an alternating circuit varies decisively. First, however, we will look at a component whose behaviour is constant. We are talking about an ohmic resistor.

14.1 The resistance

A resistor in the circuit is an obstacle to the current. This also applies in an alternating current circuit. The simplest circuit consists of an AC voltage source and a simple resistor.

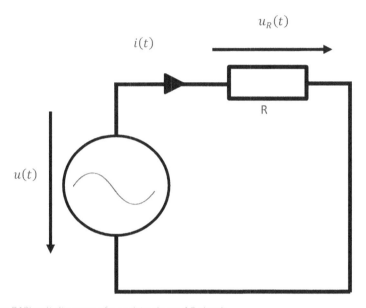

Figure 74Circuit diagram of a resistor in an AC circuit

When an AC voltage is applied to an ohmic resistor, the current flows with the value at any time.

$$I = \frac{U}{R}$$

Since the AC voltage is a variable that changes over time, the current curve follows the voltage curve.

$$i(t) = \frac{\hat{U}}{R} \times \sin(2\pi \times f \times t) = \hat{I} \times \sin(2\pi \times f \times t)$$

The phase shift is therefore equal to zero. $\varphi = 0$

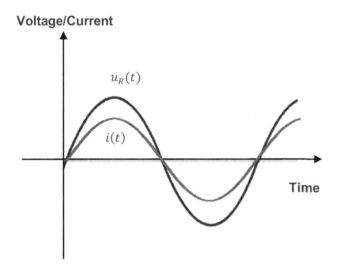

Figure 75Current and voltage curve at a resistor in an AC circuit

In practice, it is impossible to produce a perfect resistor. Every resistor, therefore, contains the minimal properties of a coil and a capacitor.

These mostly undesirable properties only occur at very high frequencies, depending on the quality of the resistor.

If one can observe such behaviour, one also speaks of parasitic effects. We will understand what this exactly means after we have looked at the behaviour of the capacitor and the coil in the AC circuit.

14.2 The capacitor

The capacitor has the ability to store electrical charges, but only for shorter periods of time.

This makes capacitors perfect for supporting fluctuating voltages and currents. The capacitor serves as a buffer store when a voltage source cannot provide the required current.

Since, in simplified terms, the capacitor merely consists of two opposing plates, it is an infinitely large resistor within a DC circuit. After all, the electrons cannot "jump" from one side to the other.

However, the situation is different if we operate the capacitor in an AC circuit, for example, on a periodic sinusoidal signal.

The AC voltage increases and decreases at a constant interval. The capacitor plates are also charged and discharged in the process. The electric field between the capacitor plates also increases, reaches a peak and decreases again.

 This allows the AC voltage to be transferred from one plate to the other **completely without touching**.

An alternating current signal can be transmitted via a capacitor and its electric field. No real current flow takes place in the process. The current that is needed to build up the electric field is therefore also called **reactive current**.

The size of the capacitor or its capacity determines how quickly the landings can be shifted. Accordingly, it also determines how "well" the voltage transfer can take place.

This is why one also speaks of the capacitor representing a **reactance** in the alternating current circuit. The energy that is transferred is also called reactive energy. But what does a voltage curve actually look like when we apply a sinusoidal voltage to a capacitor? For this, we again use the example of a simple circuit, consisting of an AC voltage source and a capacitor.

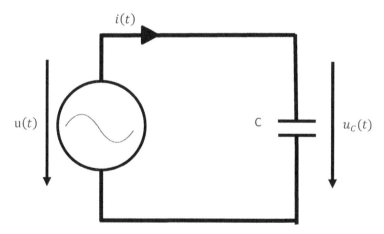

Figure 76 Circuit diagram of a capacitor in an AC circuit

The voltage curve follows that of the voltage source and is sinusoidal. But what is the current $i(t)$ that flows through the circuit?

To calculate this, we again consider when the most current can flow. This is the case when there are as few charges as possible on the plates and the potentials on the plates are very different.

At this moment, a maximum number of electrons can be moved, and as we know, current consists of moving electrons.

If the voltage subsequently remains constant, the current flow decreases and approaches zero. We have seen this behaviour, for example, when charging and discharging the capacitor.

In a capacitor, a current always flows into or out of the capacitor when the applied voltage changes. This is permanently the case in an alternating current circuit. The strength of the current depends on the **change in voltage**. For example, the voltage at the peak hardly changes. There, the current flow is also zero. In the zero crossings of the voltage, on the other hand, the voltage drops rapidly and changes its sign. There, the current flow is at a maximum.

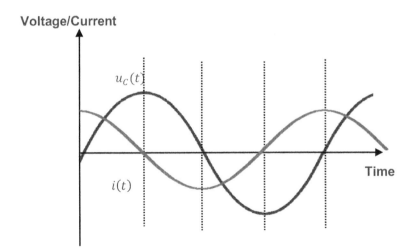

Figure 77 Current and voltage curve on a capacitor in an AC circuit

We can see the effects of the capacitive reactance on the basis of the signal curves.

This results in a current curve that is increased on the time axis by a quarter period ($\varphi = -90°$ or $\varphi = -\frac{\pi}{2}$) to the **left**.

   A popular saying in electrical engineering is:

"In the case of the capacitor, the current (voltage) **precedes**".

The phase shift is always the same. However, this reactance is not constant. It depends on two things: Firstly, on the capacitance of the capacitor. The higher the capacitance, the easier it is for new charge carriers to reach the plates of the capacitor and the easier it is to charge or discharge it. A larger capacitance produces a lower reactance.

Secondly, the resistance depends on the rate of change of the voltage wave.

A slowly changing sine wave can be transmitted much worse than one that changes quickly. We already know the unit used to denote how fast or how often the sine wave changes in a period. It is the frequency f.

This brings us to a frequency-dependent reactance. It is abbreviated with the symbol X_C and can be calculated as:

$$X_C = -\frac{1}{\omega \times C} = -\frac{1}{2\pi f \times C}$$

The minus sign results from the fact that the shift of the current in relation to the voltage is negative. For purely effective value considerations, the minus sign is usually negligible. It is therefore often omitted for the sake of simplicity.

$$X_C = \frac{1}{\omega \times C} = \frac{1}{2\pi f \times C}$$

Since it is a resistor, the unit of reactance is also the ohm Ω.

It is also very difficult to determine the instantaneous current or the active power of a capacitor. Therefore, we use the effective values of the sinusoidal voltages. These are related via "Ohm's law of the alternating current circuit".

$$U_{Eff} = X_C \times I_{Eff}$$

We can see why we have become familiar with the RMS values of the voltages. This allows us to use similar formulas in the AC circuit as in the DC circuit.

A sinusoidal voltage is connected to a capacitor with a capacitance of 1 μF at a frequency of $f = 100\ Hz$ is connected. What is the reactance of the capacitor in this case?

Components in the AC circuit

Solution:
$$X_C = \frac{1}{\omega \times C} = \frac{1}{2\pi \times 100\ Hz \times 1 \times 10^{-6}F} = 1591\ \Omega$$

What is the reactance of a capacitor with a capacity of 47 µF when we connect it to the low-voltage mains ($U_{Eff} = 230\ V$, $f = 50\ Hz$)? What is the RMS value of the current flowing through the capacitor?

$$X_C = \frac{1}{\omega \times C} = \frac{1}{2\pi \times 50\ Hz \times 47 \times 10^{-6}F} = 67.73\ \Omega$$

$$U_{Eff} = X_C \times I_{Eff}$$

$$I_{Eff} = \frac{U_{Eff}}{X_C} = \frac{230\ V}{67.73\ \Omega} = 3.4\ A$$

Even though the formulae for the AC circuit are very similar to those of the DC circuit, we must always bear in mind that we are calculating with effective values and not with peak values.

One example of this can be seen in our electricity grid. This has an effective voltage of $U_{Eff} = 230\ V$. When a power supply manufacturer designs a circuit to be connected to the mains, he must bear in mind that the peak value of the voltage is decisive. It is **not** enough for him to buy a capacitor with a maximum dielectric strength of, for example, 250 V. Because the peak value of the sinusoidal voltage is $U_{Eff} \times \sqrt{2}$, i.e., approximately 325 V. A capacitor with a maximum dielectric strength of 250 V would explode immediately during commissioning.

14.3 The coil

In the AC circuit, the coil also has many similarities with the capacitor.

A current-carrying coil has the ability to build up a magnetic field and thus store electrical energy. This is why coils are used in transformers, for example.

Another aspect is the self-induction of the coil. This prevents the rapid increase of a current within the coil, for example during the switching-on and switching-off process of a coil.

Self-induction slows down the build-up of the current. Therefore, self-induction creates a resistance for the current – a reactance.

In the case of the coil, this is referred to as an inductive reactance.

In the capacitor, the reactance is caused by the electric field between the plates, in the coil by the magnetic field during self-induction.

When a voltage is applied, the electrons want to flow through the wire of the coil but are slowed down by the magnetic field that builds up.

With alternating current, this effect is intensified because the alternating current constantly builds up and dissipates a magnetic field in the coil. When the field is built up, the coil absorbs and stores energy. When the magnetic field is reduced, the coil releases the energy again.

With the coil, too, we look at the current curve when we apply a sinusoidal voltage to the coil. For this, we again use a simple circuit consisting of an AC voltage source and a coil.

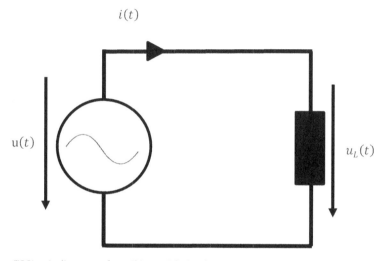

Figure 78Circuit diagram of a coil in an AC circuit

Analogous to the procedure with the capacitor, we consider when the most current can flow. Since the build-up of the magnetic field requires energy, the current in the coil is delayed in relation to the voltage.

The most current can flow in the coil when the voltage changes strongly. This is the case in the zero crossings of the sinusoidal voltage curve. There, no more energy is needed to build up the magnetic field. At this point, it is at a maximum.

On the other hand, when the voltage reaches a peak value, the change and the current are almost zero.

Components in the AC circuit

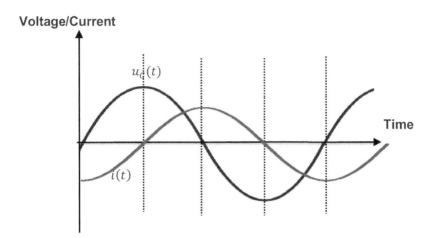

Figure 79 Current and voltage curve on a coil in an AC circuit

We can see the effects of the inductive reactance on the basis of the signal curves.

This results in a current curve that is increased by a quarter period ($\varphi = 90°$ or $\varphi = \frac{\pi}{2}$) to the **right**.

 The analogous mnemonic here for the coil is:
"With inductance, the current comes too **late**. "

The inductive reactance is also not constant. It depends on two things. Firstly, on the inductance of the coil. The higher the (self-) inductance of the coil, the larger the magnetic field that is created and the more energy is needed to build it up. A greater inductance produces a greater reactance.

Secondly, the resistance depends on the frequency of the applied sinusoidal voltage.

 A rapidly changing sine wave builds up a magnetic field more frequently and can therefore be transmitted much less effectively than a sine wave with a low frequency.

This brings us to a frequency-dependent reactance. It is abbreviated with the symbol X_L and can be calculated as:

$$X_L = \omega \times L = 2\pi f \times L$$

All other properties and formulae relating to the reactance are identical. Thus, the unit of inductive reactance is also the ohm Ω. We can also apply "Ohm's law of alternating current theory".

$$U_{Eff} = X_L \times I_{Eff}$$

 A sinusoidal voltage with a frequency of $f = 100\ kHz$ is connected. What is the reactance of the coil in this case?

Solution:
$$X_L = \omega \times L = 2\pi \times 1 \times 10^5 Hz \times 1 \times 10^{-6}\ H = 0.63\ \Omega$$

 What is the reactance of a coil with a capacitance of 47 mH that we connect to the low-voltage mains ($U_{Eff} = 230\ V, f = 50\ Hz$)? What is the RMS value of the current that flows through the capacitor?

$$X_L = \omega \times L = 2\pi \times 50\ Hz \times 470 \times 10^{-3}\ H = 14.8\ \Omega$$
$$U_{Eff} = X_C \times I_{Eff}$$
$$I_{Eff} = \frac{U_{Eff}}{X_L} = \frac{230\ V}{14.8\ \Omega} = 15.5\ A$$

Here, too, we must always bear in mind that we are calculating with effective values and not with peak values.

 In the last example, we get an effective current $I_{Eff} = 15.5\ A$. However, the peak value of the sinusoidal voltage is $I_{Eff} \times \sqrt{2}$, i.e., approx. 22 A.

A fuse that can withstand a maximum of 20 A at its peak would therefore blow.

 However, a standard household 16 A fuse would most likely not blow, as these are standardised to the RMS value and not the peak value.

Strictly speaking, fuses are triggered by heating or different expansion of a bimetal. However, the heating is only dependent on the converted power, i.e., the effective value.

Now that we are familiar with reactance, let's look at different types of power that result from the displacement of current and voltage.

Components in the AC circuit

14.4 Active, reactive and apparent power

A quantity that seems quite obscure at first glance is the **reactive power Q**.

The reactive power is the analogue of the "normal" power P in the DC circuit. To separate the types of power, the power P is also called active power.

The active power is the power that acts on the components, e.g., that heats up a resistor. Its unit is the watt. It occurs when current and voltage **act simultaneously**.

$p(t) = u(t) \, x \, i(t)$

Here, too, we need an extension for AC voltage quantities. The relationship of the active power results from the effective values of current and voltage as well as the relative phase shift and is calculated as follows:

$P = U_{Eff} \, x \, I_{Eff} \, x \, \cos \varphi$

Reactive power, on the other hand, is the power that results from the phase shift between current and voltage. Reactive power, therefore, does not make a motor turn, an LED light up, or a heating element heat up.

This is a pure **current shift**, for example, to charge a capacitor or a coil.

The reactive power is abbreviated with the quantity Q. The unit "watts" was deliberately avoided to distinguish it from active power. Instead, the unit for reactive power is volt-ampere-reactive, or **var** for short.

The reactive power can be calculated using the RMS values of current and voltage and the relative phase shift:

$Q = U_{Eff} \, x \, I_{Eff} \, x \, \sin \varphi$

We must note that the formula sign Q is also used for the charge. Therefore, the context in which we use the formula sign is important.

Mostly, reactive power is a side effect that occurs when charging and discharging capacitors and coils. For example, grid operators must always make sure that the reactive power in the power grid is compensated. Otherwise, voltage surges or voltage dips take place. These can occur when, for example, large inductive machines are switched on. In response, the grid operators have to add huge coils or capacitors to counteract this.

Figure 80Compensation coil in the power grid for power factor correction

Source: https://www.dehn-international.com/en/node/1252

As the name suggests, only reactive power occurs with an inductive (coil) or capacitive (capacitor) reactance. With a phase shift of current and voltage of exactly 90°, **only** reactive power occurs. This is also confirmed by the formulas we have learned:

The following applies to the coil and capacitor:

$P = U_{Eff} \times I_{Eff} \times \cos \varphi = U_{Eff} \times I_{Eff} \times \cos \pm 90° = U_{Eff} \times I_{Eff} \times 0 = \mathbf{0}$

$Q = U_{Eff} \times I_{Eff} \times \sin \varphi = U_{Eff} \times I_{Eff} \times \sin \pm 90° = U_{Eff} \times I_{Eff} \times \pm 1$

$= \pm U_{Eff} \times I_{Eff}$

On the other hand, the following applies to the ideal ohmic resistance, which contains no reactance components:

$P = U_{Eff} \times I_{Eff}$

$Q = 0$

Another power parameter that combines the two types of power, active and reactive power, is the **apparent power S.**

In an AC circuit, the apparent power is the product of the effective value of the voltage and the current. The unit is accordingly **volt-ampere VA**. In the DC circuit, the volt-ampere is equal to the watt. In the AC circuit, the units are deliberately used differently.

$S = U_{Eff} \times I_{Eff}$

Components in the AC circuit

Apparent power, active power and reactive power are related via a power triangle.

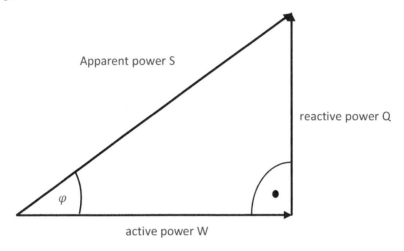

Figure 81 Power triangle in the AC circuit

Now we also understand the formulas and relationships of the three types of power. This is the application of the Pythagorean theorem or the cosine and sine.

$$S^2 = W^2 + Q^2$$

$$\sin \varphi = \frac{Q}{S} \rightarrow Q = S \times \sin \varphi = U_{Eff} \times I_{Eff} \times \sin \varphi$$

$$\cos \varphi = \frac{W}{S} \rightarrow W = S \times \cos \varphi = U_{Eff} \times I_{Eff} \times \cos \varphi$$

In practical applications, such as the European power grid already mentioned, the aim is to achieve a high proportion of active power and a low proportion of reactive power. Accordingly, $\cos \varphi$ should be close to one. Therefore, the factor $\cos \varphi$ is also described as the **active factor**. It indicates the ratio of active to apparent power and is often given as a percentage.

An induction cooker is fed from the public power grid. ($U_{Eff} = 230\ V, f = 50\ Hz$) mains. It draws an active power of $P = 4\ kW$ and an inductive reactive power of $Q = 3\ kVA$ is consumed.

What is the apparent power S?

What is the RMS current I_{Eff} ?

How high is the impact factor?

By what angle are current and voltage shifted?

How could one increase the effective factor (compensate for reactive power)?

Solution:

$S^2 = W^2 + Q^2$

$S = \sqrt{W^2 + Q^2} = \sqrt{(4\ kW)^2 + (3\ kvar)^2} = 5\ kVA$

$S = U_{Eff} \times I_{Eff} \rightarrow I_{Eff} = \dfrac{S}{U_{Eff}} = \dfrac{5\ kVA}{230\ V} = 21{,}74\ A$

The impact factor is there: $\cos \varphi = \dfrac{W}{S} = \dfrac{4\ kW}{5\ kVA} = 0.8 = 80\%$

Displacement angle: $\cos \varphi = 0.8 \rightarrow arccos(0.8) = 36{,}9°$

(Setup calculator to **Degree**, **not Radian**)

To compensate for inductive reactive power, capacitive reactive power must be added. One could therefore connect a capacitor in parallel to the induction cooker.

Components in the AC circuit

14.5 The electromagnetic oscillating circuit

Lastly, let's look at a commonly used circuit that allows us to use any type of wireless transmission.

The circuit is an **LC resonant circuit**. Some of you may have already guessed that the LC resonant circuit is a circuit consisting of a coil (L) and a capacitor (C).

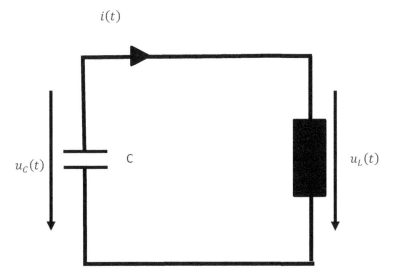

Figure 82Circuit diagram of an electromagnetic oscillating circuit

This circuit consists exclusively of passive elements, namely a coil and a capacitor. One looks in vain for a voltage source.

In order for the circuit to start oscillating, energy must be supplied to the circuit. This is possible, for example, by applying an electric or magnetic field from the outside for a short moment.

In this example, we assume the following initial states:

- The capacitor is fully charged. The E-field is, therefore, at a maximum, and the maximum voltage is applied across the capacitor.

- The current in the circuit is zero. The coil has, therefore, not built up a magnetic field, and the voltage across the coil is equal in magnitude to the voltage across the capacitor.

Analogously, we could also consider what would happen if the coil were charged first. Likewise, all intermediate stages are completely equivalent. For this example, however, only the capacitor is fully charged at the time of the $t = 0$.

After energy has been supplied to the circuit, the capacitor wants to discharge. (see 10.2Discharging the capacitor).

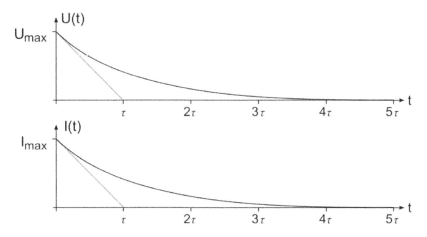

Figure 83Discharge curves of a capacitor

We recall that in this case, the initial current is maximum and gradually decays. However, this is not possible with the circuit because the coil prevents a sudden increase in the current at time $t = 0$. Instead, the current can only rise slowly. This results in a current-voltage curve, which is shown in the following figure.

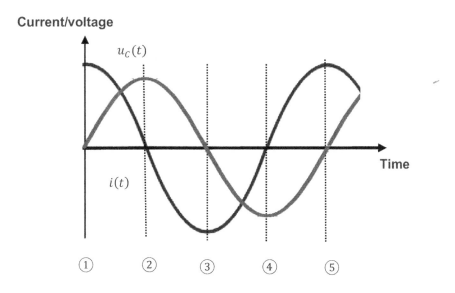

Figure 84Current and voltage curve in an electromagnetic oscillating circuit

Components in the AC circuit

① The energy stored in the electric field of the capacitor is dissipated. At the same time, the current and the resulting magnetic field around the coil increase. The energy flows from the electric field into the magnetic field.

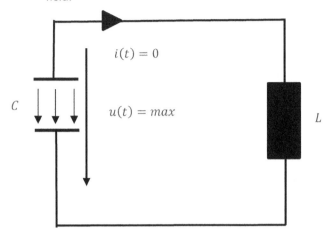

Figure 85 Initial state of the electromagnetic oscillation

② After a time, the capacitor is completely discharged. The current flow and the magnetic field are at a maximum.

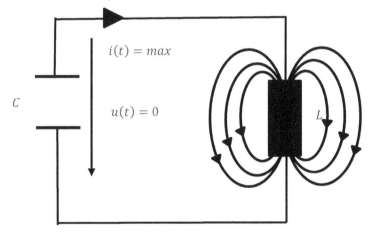

Figure 86 The magnetic field is fully formed

However, the inductance of the coil now also prevents the current from stopping abruptly. We remember that the coil, like a large paddle wheel, ensures that the current continues to flow. As a consequence, the current slowly decreases, negatively charging the capacitor. At the

③ same time, the magnetic field around the coil decreases. The energy is used to build up the electric field.

After the current has completely decayed and the current of the circuit has become zero, the whole process starts again. This time, the capacitor's polarity is reversed, so the voltage is maximum negative.

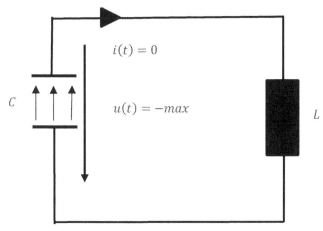

Figure 92 The capacitor is inversely charged

The capacitor then discharges again, the current changes direction and increases. The magnetic field also forms again. Since the current flows in the other direction, the current in our arrow system is negative, as is the resulting B field.

④ Analogous to stage ②, the capacitor is completely discharged. The voltage is accordingly equal to zero. The current flow and the magnetic field are at a maximum. The current charges the capacitor again while the magnetic field is reduced.

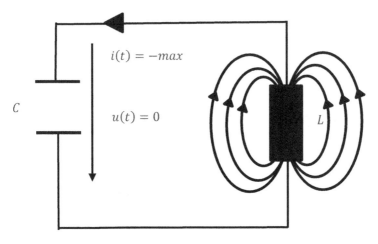

Figure 87 The current flow changes direction

⑤ The initial state is reached again. From here on, the process repeats itself. A harmonic oscillation is created.

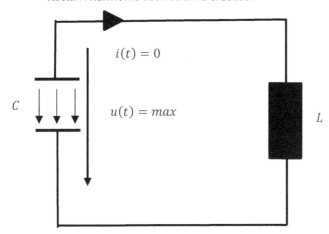

Figure 88 The initial state is reached again

This is why the circuit is also called an **LC resonant circuit** or **electromagnetic resonant circuit** – the energy is stored alternately by the electric and magnetic field.

We can calculate the frequency of the oscillation, i.e., how fast the energy can be "pushed" back and forth between the components, using the parameters of the capacitor and coil.

Derivation

In order to derive the frequency at which the oscillating circuit oscillates, we look at the resistance of the circuit. The resistance of the entire circuit results from the series connection of the resistances of the capacitor and coil. The resistances are added together.

$$X_C = \frac{1}{\omega \, x \, C} \, ; X_L = \omega \, x \, L$$

$$X_{LC} = X_L + X_C = \omega \, x \, L - \frac{1}{\omega \, x \, C}$$

In this context, note the minus sign of the capacitive reactance. The capacitor shifts the current forward (minus) by a quarter period compared to the voltage, and the coil shifts it backward (plus).

The resulting total resistance should become minimal so that the oscillation can propagate as free of damping as possible. This happens exactly at the point when the inductive reactance of the coil and the capacitive reactance of the capacitor cancel each other out. The resulting total resistance is zero.

$$X_L = X_C => X_{LC} = \omega \, x \, L - \frac{1}{\omega \, x \, C} = 0$$

Using this equation, we can determine the frequency or the angular frequency. By rearranging the equation according to the quantity we are looking for, we obtain the following:

$$\omega \, x \, L - \frac{1}{\omega \, x \, C} = 0$$

$$\omega \, x \, L = \frac{1}{\omega \, x \, C}$$

$$\omega^2 = \frac{1}{L \, x \, C}$$

$$\Rightarrow \omega = \sqrt{\frac{1}{L \, x \, C}} \quad bzw. f = \frac{1}{2\pi} \, x \, \sqrt{\frac{1}{L \, x \, C}}$$

The frequency of an LC resonant circuit is also called the **resonant frequency** f_0. At this frequency, the oscillating circuit is in resonance.

This relationship was discovered by the British physicist William Thomson in 1853. With the help of Thomson's oscillation equation, the resonant frequency of an LC resonant circuit can be determined.

We have derived the formula for determining the resonant frequency using a series resonant circuit. However, the formula also applies to a parallel resonant circuit. In this case, the coil and capacitor are connected in parallel instead of in series.

Components in the AC circuit

An LC resonant circuit consisting of a capacitor with a capacitance of $C = 22\ \mu F$ and a coil with an inductance of $L = 470\ \mu H$ is excited. How often does the voltage oscillate per second? How often does the current oscillate? What is the capacitive reactance of the capacitor and the inductive reactance of the coil?

Solution: $f = \dfrac{1}{2\pi} \times \sqrt{\dfrac{1}{L \times C}} = \dfrac{1}{2\pi} \times \sqrt{\dfrac{1}{470\ \mu H \times 22\ \mu H}} = 1.565\ kHz$

Both the current and the voltage oscillate at 1565 oscillations per second.

$X_L = X_C = \omega \times L = 2\pi \times 1.565\ kHz \times 470\ \mu H = 4.6\ \Omega$

After learning how an electromagnetic oscillation is created, we look at how it can be used for wireless data transmission.

14.6 Electromagnetic radiation

An electromagnetic oscillation or an electromagnetic wave is created by the interaction of electric and magnetic fields, for example, with the help of an LC resonant circuit.

At low frequencies, the electrons oscillate back and forth in the form of a measurable voltage and current in the conductors or cables.

However, if we increase the frequency at which the electrons oscillate back and forth and add a suitable antenna, the oscillations can detach from the conductive paths into space.

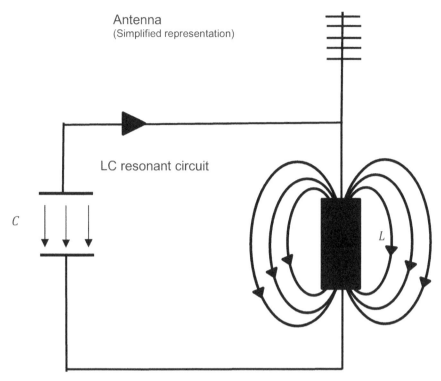

Figure 89 Set-up of an electromagnetic oscillating circuit with output antenna

The wave is **radiated** into the room. To be able to achieve this effect, the frequencies chosen must be extremely high.

 When an electromagnetic wave is radiated from the conductor, the oscillating electrons remain in the conductor because a wave does not transport any matter. Only the electrical energy is converted into radiant energy by the wave and radiated.

 The electric and magnetic fields are spatially shifted by 90° and are in turn 90° to the propagation speed.

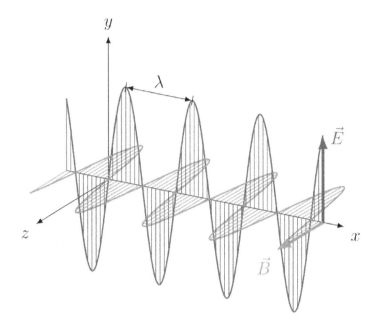

Figure 90 Illustration of the propagation of an electromagnetic wave in space

It has already been mentioned that the frequency of the oscillation chosen must be very high so that the wave propagates. The following table helps us to get a feel for the magnitudes of different waves. Since the waves are classified according to their frequencies, we also speak of a (frequency) spectrum.

Designation	Wave frequency	Example
Low frequency	0 Hz to 50 Hz	---
	50 Hz	European electricity grid (conductor-based)
	Up to 30 kHz	Submarine communication
High frequency	Up to 3 MHz	Shortwave radio
	Up to 300 MHz	Radio and TV
	Up to 1 GHz	Mobile radio
	2.4 GHz	2.4-WLAN
	Up to 5 GHz	Bluetooth, 5G, GPS
	Up to 80 GHz	Radar
Infrared (heat radiation)	> 300 GHz	Microwaves Radiant heater
Light	> 300 THz	Visible light
UV rays	> 800 THz	Black light, photolithography
X-rays	> 30,000 THz	Medical technology

The European power grid is operated at a very low frequency of 50 Hz. This is because it is a **wired oscillation** that is **not** supposed to detach. Every electromagnetic wave that detaches from the cables means a **loss of energy** during transport.

15 Conclusion

In the meantime, we have learned the basics of electrical engineering, dealt with various components and explained the corresponding circuit symbols. Furthermore, we have dealt with the water model and the basic electrotechnical quantities and effects.

Finally, we have gained an understanding of simple circuits such as the charging and discharging of the coil and capacitor.

We have also looked at more complex topics, such as AC theory and LC resonant circuits including practical examples such as the structure of the power grid.

Of course, this was just an introduction to get familiar with the subject. The world of electrical engineering is much more extensive. More complex components exist, like ICs (integrated circuits), which contain entire circuits within themselves.

Important basic knowledge, such as the different sizes of current, voltage, resistance and much more, is now known, as are the units for different types of power or energy.

According to the motto, "No master has fallen from the sky", it is important to apply what you have learned and not to stand still. Because Nikola Tesla already knew:

"The progressive development of mankind depends in a vital way on invention."

-Nikola Tesla

Free eBook

Thanks for buying this book. Since the formatting of the book is done directly by Amazon and I have no influence on the quality of the pictures, it is possible that details may be lost or the formatting may suffer.

That's why I offer the eBook as a PDF file free of charge when you buy the book. There all pictures are high resolution, the formatting fits and you always get the latest version.

To claim this, send a message with the subject "electrical engineering eBook ", as well as a screenshot of the purchase or proof of order from Amazon to the email:

BenjaminSpahic@pbd-verlag.de

I will send you the eBook immediately.

The book was translated and proofread. Nevertheless, there could be errors. Also, translation errors!

If you are missing something, did not like something or you have suggestions for improvement or questions, please send me an e-mail.

Constructive criticism is important to be able to improve something. I am constantly revising the book and am happy to respond to any constructive suggestions for improvement.

Otherwise, if you liked the book, I would also appreciate a positive review on Amazon. This helps the visibility of the book and is the biggest praise an author can get.

Yours, Benjamin

About the author

Benjamin Spahic was born in Heidelberg, Germany in 1995 and grew up in a village of 8,000 souls near Karlsruhe. His passion for technology is reflected in his studies of electrical engineering with focus on information technology at the University of Applied Sciences in Karlsruhe.

Afterwards, he deepened his knowledge in the field of regenerative energy production at the Karlsruhe University of Applied Sciences.

Picture credits:
https://icons8.de/icon/113140/kugelbirne
https://icons8.de/icon/79638/obligatorische
https://icons8.de/icon/78038/math
https://icons8.de/icon/42314/taschenrechner
All non-mentioned contents were created by the author himself. He is therefore the author of the graphics and has the rights of use and distribution.
https://pixabay.com/de/illustrations/atom-molek%C3%BCl-wasserstoff-chemie-2222965/
https://en.wikipedia.org/wiki/File:Electrostatic_induction.svg
https://commons.wikimedia.org/wiki/File:Feldlinien_und_%C3%84quipotentiallinien.png
*: https://commons.wikimedia.org/wiki/File:VFPt_cylindrical_magnet_thumb.svg
*: https://de.wikipedia.org/wiki/Datei:RechteHand.png
*: https://de.wikipedia.org/wiki/Datei:Lorentzkraft_v2.svg
https://commons.wikimedia.org/wiki/File:RHR.svg
https://de.wikipedia.org/wiki/Datei:Schaltzeichen_Masse.svg
https://de.wikipedia.org/wiki/Datei:Chassis_Ground.svg
https://commons.wikimedia.org/wiki/File:Stromknoten.svg
*: https://de.wikipedia.org/wiki/Datei:Widerst%C3%A4nde.JPG
*: https://commons.wikimedia.org/wiki/File:Manta_DVD-012_Emperor_Recorder_-_power_supply.JPG
https://de.wikipedia.org/wiki/Datei:Diodenalt2.png
*: https://de.wikipedia.org/wiki/Datei:Diode_pinout_de.svg
**: https://www.chemie-schule.de/KnowHow/Datei:Sperrschicht.svg
**: https://www.chemie-schule.de/KnowHow/Datei:Sperrschicht.svg
https://de.wikipedia.org/wiki/Datei:Transistors-white.jpg
https://upload.wikimedia.org/wikipedia/commons/3/3f/Dreiphasenwechselstrom.svg
https://commons.wikimedia.org/wiki/File:Transistor-diode-npn-pnp.svg
https://de.wikipedia.org/wiki/Datei:NPN_transistor_basic_operation.svg
*: https://de.wikipedia.org/wiki/Datei:N-Kanal-MOSFET_(Schema).svg
**: https://de.wikipedia.org/wiki/Datei:Scheme_of_metal_oxide_semiconductor_field-effect_transistor.svg
**:https://de.wikipedia.org/wiki/Datei:Scheme_of_n-metal_oxide_semiconductor_field-effect_transistor_with_channel_de.svg
*: https://commons.wikimedia.org/wiki/File:MISFET-Transistor_Symbole.svg
*: https://de.wikipedia.org/wiki/Datei:Elko-Al-Ta-Bauformen-Wiki-07-02-11.jpg
*: https://commons.wikimedia.org/wiki/File:Kondensatoren-Schaltzeichen-Reihe.svg
*: https://de.wikipedia.org/wiki/Datei:Plate_Capacitor_DE.svg
https://de.wikipedia.org/wiki/Datei:Ladevorgang.svg
https://de.wikipedia.org/wiki/Datei:Ladevorgang.svg
*: https://de.wikipedia.org/wiki/Datei:Electronic_component_inductors.jpg
https://de.wikipedia.org/wiki/Datei:Diverse_Spulen.JPG
https://commons.wikimedia.org/wiki/File:Solenoid-1.png
https://de.wikipedia.org/wiki/Datei:Ladevorgang.svg
https://de.wikipedia.org/wiki/Datei:Ladevorgang.svg
https://www.flaticon.com/de/premium-icon/strommast_3573229?term=strommast&page=1&position=12&page=1&position=12&related_id=3573229&origin=search
https://www.flaticon.com/de/premium-icon/sonnenkollektor_3933850
https://www.flaticon.com/de/premium-icon/okonach-hause_4640172
https://www.flaticon.com/de/kostenloses-icon/windkraft_902587
https://www.flaticon.com/de/premium-icon/wasserkraft_3202537
https://www.flaticon.com/de/kostenloses-icon/wasserkraft_259011

* This file is made available under the GNU Free Documentation License.
https://commons.wikimedia.org/wiki/Commons:GNU_Free_Documentation_License,_version_1.2
Changes may have been made.
** This file is made available under the Creative Commons licence "CC0 1.0 waiver of copyright".
https://creativecommons.org/publicdomain/zero/1.0/deed.de
Changes may have been made.

Index

AC 115
active factor 145
active power 143, 145
air coil 95, 96
alternate current 115
alternating current source
 133
alternating voltage 115
amplitude 116
angular frequency........116
apparent power 144
Apparent power 145
arc cosinee 17
arc sine 17
arc tangent.....................17
backup capacitors 93
barrier layer 73
Blindleistung 145
capacitance 85
capacitor 83, 135
carrier material...............72
cloud-to-earth lightning 109
coil 95, 139
concatenation factor 130
Conductance 65
conductor-to-conductor
  voltage 130
consumer arrow system 59
current divider................68
current restriction...........98
current shift..................143
cylindrical coil 96
DC.................................109
depletion zone 73
differential 49
differential equation 88
diffuse 73
diode 71
direct current................109
doped.............................72
Drain 79
Drehstrom....................127
earth..............................55
effective value..............119
electric field constant.....85

Electric potential............ 39
electrical generators ... 112
Electrical resistance...... 63
electricity grid 124
electrodes 84
electromagnetic induction
  48
elementary magnets 44
energy 28
equipotential lines 39
equivalent transformations
  8
Euler's number e 11
excitation coils............. 123
Exponential functions...... 9
ferrite 95
FET 79
field effect transistor...... 79
filter 93
foreign atom 72
frequency 116
Gate 79
generator arrow system 59
GND 55
hypotenuse 16
inductance.................... 96
inductive reactance 141
Kirchhoff's laws 59
law of induction 49
LC resonant 151
LC resonant circuit 147
Lenz's rule 50
Lorentz force 51
maximum voltage........ 116
meshes.......................... 61
MOSFET 81
n-doped 72
node theorem................ 60
NPN-transistor 77
oxidises 110
parallel connection 67
parasitic effects 135
p-doped 72
peak value.................. 116
period 116

permeability 43
phase...........................126
phase difference..........126
phase position 126
phase shift 126
physical direction of the
  current 42
p-n-junction 73
PNP-transistor 77
potential difference 41
potential of the earth......55
power 28, 69
power current...............127
power triangle 145
prefixes 22
protective conductor 127
reactance.....................136
reactive current............136
reactive power.....143, 145
recombination 73
relative permittivity.........85
RL element 99
RMS-value 119
root mean square value
 119
Root-Mean-Square......119
sample charge 51
self-inductance 96
short circuit 57
solar cell 112
solar panel 112
Source 79
technical direction of
  current 42
time average value 117
time constant .89, 101, 104
trigonometry..................15
Volta cells 110
Volta Column 110
voltage divider 66
volt-ampere..................144
work 28
zero point....................127
zero potential 55

Free eBook 155

Made in United States
North Haven, CT
01 August 2023

39782761R00095